# Julia Quick Syntax Reference

## A Pocket Guide for Data Science Programming

## Second Edition

Antonello Lobianco

Apress®

*Julia Quick Syntax Reference: A Pocket Guide for Data Science Programming, Second Edition*

Antonello Lobianco
Nancy, France

ISBN-13 (pbk): 979-8-8688-0964-4        ISBN-13 (electronic): 979-8-8688-0965-1
https://doi.org/10.1007/979-8-8688-0965-1

Copyright © 2024 by Antonello Lobianco

This work is subject to copyright. All rights are reserved by the Publisher, whether the whole or part of the material is concerned, specifically the rights of translation, reprinting, reuse of illustrations, recitation, broadcasting, reproduction on microfilms or in any other physical way, and transmission or information storage and retrieval, electronic adaptation, computer software, or by similar or dissimilar methodology now known or hereafter developed.

Trademarked names, logos, and images may appear in this book. Rather than use a trademark symbol with every occurrence of a trademarked name, logo, or image we use the names, logos, and images only in an editorial fashion and to the benefit of the trademark owner, with no intention of infringement of the trademark.

The use in this publication of trade names, trademarks, service marks, and similar terms, even if they are not identified as such, is not to be taken as an expression of opinion as to whether or not they are subject to proprietary rights.

While the advice and information in this book are believed to be true and accurate at the date of publication, neither the authors nor the editors nor the publisher can accept any legal responsibility for any errors or omissions that may be made. The publisher makes no warranty, express or implied, with respect to the material contained herein.

> Managing Director, Apress Media LLC: Welmoed Spahr
> Acquisitions Editor: Melissa Duffy
> Development Editor: James Markham
> Editorial Assistant: Gryffin Winkler

Cover designed by eStudioCalamar

Distributed to the book trade worldwide by Springer Science+Business Media New York, 1 New York Plaza, Suite 4600, New York, NY 10004-1562, USA. Phone 1-800-SPRINGER, fax (201) 348-4505, e-mail orders-ny@springer-sbm.com, or visit www.springeronline.com. Apress Media, LLC is a California LLC and the sole member (owner) is Springer Science + Business Media Finance Inc (SSBM Finance Inc). SSBM Finance Inc is a **Delaware** corporation.

For information on translations, please e-mail booktranslations@springernature.com; for reprint, paperback, or audio rights, please e-mail bookpermissions@springernature.com.

Apress titles may be purchased in bulk for academic, corporate, or promotional use. eBook versions and licenses are also available for most titles. For more information, reference our Print and eBook Bulk Sales web page at https://www.apress.com/bulk-sales.

Any source code or other supplementary material referenced by the author in this book is available to readers on GitHub. For more detailed information, please visit https://www.apress.com/gp/services/source-code.

If disposing of this product, please recycle the paper

*To my family*

to my family

# Table of Contents

About the Author .................................................................................. xv

Acknowledgments ............................................................................... xvii

Introduction ......................................................................................... xix

**Part I: Language Core** ........................................................................ 1

**Chapter 1: Getting Started** ................................................................. 3

    1.1 Why Julia ...................................................................................... 3

    1.2 Installing Julia .............................................................................. 6

    1.3 Running Julia ............................................................................... 8

    1.4 Miscellaneous Syntax Elements ................................................ 10

    1.5 Modules and Packages .............................................................. 11

        1.5.1 Using the Package Manager ............................................ 12

        1.5.2 Using Modules ................................................................. 14

        1.5.3 Using Packages ............................................................... 16

    1.6 Environments ............................................................................. 17

    1.7 Help System ............................................................................... 21

**Chapter 2: Data Types and Structures** ............................................. 23

    2.1 Simple Types (Non-Containers) ................................................. 24

        2.1.1 Basic Mathematical Operations ....................................... 25

    2.2 Strings ........................................................................................ 25

        2.2.1 Concatenation .................................................................. 26

TABLE OF CONTENTS

 2.3 Arrays (Lists) ............................................................................27
  2.3.1 Multidimensional and Nested Arrays ..................................32
 2.4 Tuples ...................................................................................... 37
 2.5 Named Tuples ......................................................................... 39
 2.6 Dictionaries .............................................................................. 39
 2.7 Sets .......................................................................................... 41
 2.8 Dates and Times ...................................................................... 42
  2.8.1 Creation of a Date or Time Object ("Input") ....................... 42
  2.8.2 Extraction of Information from a Date/Time Object ("Output") ............. 44
  2.8.3 Periods and Date/Time Arithmetic ...................................... 45
 2.9 Memory and Copy Issues ........................................................ 47
 2.10 Random Numbers .................................................................. 50
 2.11 Missing, Nothing, and NaN .................................................... 52
 2.12 Various Notes on Data Types ................................................ 53
  2.12.1 Variable References ......................................................... 54

**Chapter 3: Control Flow and Functions ......................................... 55**
 3.1 Code Block Structure and Variable Scope .............................. 55
 3.2 Repeated Iteration: for and while Loops, List Comprehension, Maps ... 57
 3.3 Conditional Statements: if Blocks, Ternary Operator .............. 60
 3.4 Functions ................................................................................. 62
  3.4.1 Arguments .......................................................................... 63
  3.4.2 Return Value ...................................................................... 66
  3.4.3 Multiple-Dispatch ............................................................... 66
  3.4.4 Templates (Type Parameterization) ................................... 67
  3.4.5 Functions As Objects ......................................................... 67
  3.4.6 Call by Reference/Call by Value ........................................ 68

TABLE OF CONTENTS

    3.4.7 Anonymous Functions (a.k.a. "Lambda" Functions) .......................... 69

    3.4.8 Broadcasting Functions ............................................................. 70

  3.5 do Blocks ............................................................................................. 71

  3.6 Exiting Julia ......................................................................................... 72

## Chapter 4: Custom Types ............................................................. 73

  4.1 Primitive Type Definition ..................................................................... 74

  4.2 Structure Definition ............................................................................. 75

  4.3 Object Initialization and Usage ............................................................ 76

  4.4 Abstract Types and Inheritance ........................................................... 79

    4.4.1 Implementation of the Object-Oriented Paradigm in Julia .................. 82

  4.5 Some Useful Functions Related to Types ............................................. 85

  4.6 Definition of Common Julia Terms ....................................................... 86

  4.7 Exercise 1: The Schelling Segregation Model ....................................... 87

    4.7.1 Instructions ................................................................................ 89

    4.7.2 Skeleton .................................................................................... 89

    4.7.3 Results ...................................................................................... 95

    4.7.4 Possible Variations ..................................................................... 95

## Chapter 5: Input/Output ............................................................. 97

  5.1 File System Functions ......................................................................... 98

  5.2 Reading (Input) ................................................................................. 100

    5.2.1 Reading from a Terminal ........................................................... 100

    5.2.2 Reading from a File .................................................................. 101

    5.2.3 Importing Data from Excel ........................................................ 104

    5.2.4 Importing Data from JSON ........................................................ 104

    5.2.5 Accessing Web Resources ........................................................ 106

  5.3 Writing (Output) ................................................................................ 108

    5.3.1 Writing to the Terminal ............................................................. 108

## TABLE OF CONTENTS

5.3.2 Writing to a File ..................................................................110

5.3.3 Exporting to CSV ...............................................................111

5.3.4 Exporting to Excel and OpenDocument Spreadsheet (ODS) Files .......112

5.3.5 Exporting to JSON ..............................................................114

5.4 Other Specialized IO ...............................................................114

### Chapter 6: Metaprogramming and Macros ..................................115

6.1 Symbols ................................................................................116

6.2 Expressions ...........................................................................117

6.2.1 Expressions Definition ........................................................118

6.2.2 Evaluate Symbols and Expressions ....................................119

6.3 Macros ..................................................................................121

6.3.1 Macro Definition .................................................................122

6.3.2 Macro Invocation ................................................................123

6.3.3 String Macros .....................................................................124

### Chapter 7: Interfacing Julia with Other Languages ....................125

7.1 Julia ⇌ C ...............................................................................126

7.2 Julia ⇌ C++ ...........................................................................128

7.2.1 A "Hello World" Example ....................................................129

7.2.2 Passing Arguments and Retrieving Data ............................130

7.2.3 Functions with STD Classes ...............................................131

7.3 Julia ⇌ Python ......................................................................134

7.3.1 PythonCall Installation ........................................................135

7.3.2 Evaluate Python Code in Julia ............................................135

7.3.3 Use Python Packages in Julia ............................................137

7.3.4 JuliaCall (Python Package) Installation ..............................139

7.3.5 Evaluate Julia Code in Python ............................................139

7.3.6 Use Julia Packages in Python .............................................141

viii

7.4 Julia ⇌ R .................................................................................143

    7.4.1 RCall.jl Installation ..............................................................143

    7.4.2 Evaluate R Code in Julia ....................................................144

    7.4.3 Use R Packages in Julia .....................................................146

    7.4.4 JuliaCall (R Package) Installation .......................................147

    7.4.5 Evaluate Julia Code in R ....................................................148

    7.4.6 Use Julia Packages in R .....................................................150

## Chapter 8: Efficiently Write Efficient Code ............................151

8.1 Performance .............................................................................152

    8.1.1 Benchmarking ....................................................................152

    8.1.2 Profiling ..............................................................................155

    8.1.3 Type Stability .....................................................................158

    8.1.4 Other Tips to Improve Performance ...................................160

8.2 Debugging ................................................................................162

    8.2.1 Introspection Tools .............................................................163

    8.2.2 Debugging Tools ................................................................165

    8.2.3 Actions to Take Before Debugging (or While Execution Is Paused) .....167

    8.2.4 Things to Inspect when Execution Is Paused ....................168

    8.2.5 Debugging Choices when Execution Is Paused ................168

8.3 Managing Runtime Errors (Exceptions) ...................................169

## Chapter 9: Parallel Computing in Julia .................................171

9.1 GPU Programming ....................................................................172

    9.1.1 Benchmarking CPU vs. GPU ..............................................175

9.2 Multithreading (on the CPU) ....................................................176

9.3 Multiprocessing .........................................................................179

    9.3.1 Adding and Removing Processes ......................................179

    9.3.2 Running Heavy Computations on a List of Items ..............180

    9.3.3 Aggregate Results ..............................................................181

TABLE OF CONTENTS

# Part II: Packages Ecosystem ........................................................................ 183

## Chapter 10: Working with Data ............................................................... 185

### 10.1 Using the DataFrames.jl Package ........................................................ 187
#### 10.1.1 Installing and Importing the Library .............................................. 187
#### 10.1.2 Creating a DataFrame or Loading Data ........................................ 187
#### 10.1.3 Gaining Insight into the Data ........................................................ 189
#### 10.1.4 Filtering Data (Selecting or Querying Data) ................................. 190
#### 10.1.5 Editing Data .................................................................................. 193
#### 10.1.6 Editing the Structure .................................................................... 195
#### 10.1.7 Working with Categorical Data ..................................................... 197
#### 10.1.8 Managing Missing Values ............................................................ 199
#### 10.1.9 Pivoting Data ................................................................................ 202
#### 10.1.10 The Split-Apply-Combine Strategy ............................................. 206
#### 10.1.11 Dataframe Export ....................................................................... 209
### 10.2 Using IndexedTables ............................................................................ 212
#### 10.2.1 Creating an IndexedTable (`NDSParse`) ..................................... 213
#### 10.2.2 Row Filtering ................................................................................. 215
#### 10.2.3 Editing/Adding Values .................................................................. 215
### 10.3 Using the Pipe Operator ....................................................................... 216
### 10.4 Plotting .................................................................................................. 218
#### 10.4.1 Installation and Backends ............................................................ 218
#### 10.4.2 The `plot` Function ...................................................................... 220
#### 10.4.3 Plotting from Dataframes ............................................................. 224
#### 10.4.4 Plotting Densities and Distributions ............................................. 226
#### 10.4.5 Combine Multiple Plots in a Single Figure ................................... 227
#### 10.4.6 Saving a Plot ................................................................................ 228

# TABLE OF CONTENTS

## Chapter 11: Scientific Libraries ............................................. 229
### 11.1 JuMP, an Optimization Framework ........................................ 230
#### 11.1.1 The Transport Problem: A Linear Problem ........................... 232
#### 11.1.2 Choosing Between Pizzas and Sandwiches: A Nonlinear Problem ..... 241
### 11.2 SymPy, a CAS System ................................................... 244
#### 11.2.1 Loading the Library and Declaring Symbols ......................... 245
#### 11.2.2 Creating and Manipulating Expressions ............................. 246
#### 11.2.3 Solving a System of Equations .................................... 247
#### 11.2.4 Retrieving Numerical Values ....................................... 248
### 11.3 LsqFit, a Data Fit Library ............................................ 249
#### 11.3.1 Loading the Libraries and Defining the Model ...................... 249
#### 11.3.2 Parameters ....................................................... 250
#### 11.3.3 Fitting the Model ................................................. 250
#### 11.3.4 Retrieving the Parameters and Comparing the Results with the Observations ................................................... 251
### 11.4 Working with Distributions ............................................ 252
#### 11.4.1 Main Supported Distributions ...................................... 253
#### 11.4.2 API ............................................................... 255
### 11.5 EXERCISE 2: Fitting a Forest Growth Model ............................. 256
#### 11.5.1 Instructions ...................................................... 257
#### 11.5.2 Skeleton .......................................................... 257
#### 11.5.3 Results ........................................................... 264
#### 11.5.4 Possible Variations ............................................... 264

## Chapter 12: AI with Julia ................................................... 265
### 12.1 Machine Learning Goals and Approaches ................................. 266
### 12.2 The BetaML Toolkit .................................................... 268
#### 12.2.1 API and Key Principles ............................................ 271

## TABLE OF CONTENTS

**12.3 Data Preprocessing** .................................................... 275
    12.3.1 Encoding Categorical Data ................................. 275
    12.3.2 Scaling ............................................................ 277
    12.3.3 Missing Value Imputation ................................... 279
    12.3.4 Dimensionality Reduction ................................... 285
    12.3.5 Data Partitioning ............................................... 289

**12.4 Model Fitting. An Overview of the Main Algorithms** .......... 290
    12.4.1 Perceptron-Like Classifiers ................................ 290
    12.4.2 Tree-Based Models ............................................ 293
    12.4.3 Neural Networks ................................................ 295
    12.4.4 Clustering ......................................................... 305

**12.5 Model Evaluation, Interpretation, and Hyperparameter Tuning** ......... 308
    12.5.1 Regression Models ............................................ 308
    12.5.2 Classification Models ......................................... 310
    12.5.3 Clustering Models ............................................. 314
    12.5.4 Hyperparameters Evaluation ............................... 317
    12.5.5 K-fold Cross-Validation ...................................... 319
    12.5.6 Autotune ........................................................... 321
    12.5.7 Model Interpretation and Feature Importance ....... 323

**12.6 Specialized AI Libraries in Julia** ................................... 327

**12.7 EXERCISE 3: Predict the Values of Houses in Boston** ........ 329
    12.7.1 Instructions ...................................................... 331
    12.7.2 Skeleton .......................................................... 331
    12.7.3 Results ............................................................ 336
    12.7.4 Possible Variations ............................................ 336

## Chapter 13: Utilities .................................................................................. **337**

### 13.1 Weave.jl for Dynamic Documents ............................................. 337
### 13.2 ZipFile ................................................................................... 343
#### 13.2.1 Writing a Zip Archive ...................................................... 343
#### 13.2.2 Reading from a Zip Archive ............................................ 344

## Index ............................................................................................. **347**

# About the Author

**Antonello Lobianco, PhD**, is a research engineer employed by a French grande école (polytechnic university). He works on the biophysical and economic modeling of the forest sector and is responsible for the lab models portfolio. He programs in C++, Perl, PHP, Visual Basic, Python, and Julia. He teaches environmental and forest economics at the undergraduate and graduate levels and modeling at the PhD level. For a several years, Antonello has been following the development of Julia, as it fits his modeling needs. He is the author of a few Julia packages, particularly on data analysis and machine learning (search **sylvaticus** on GitHub).

# Acknowledgments

The development of this book was supported by the French National Research Agency through the Laboratory of Excellence ARBRE, part of the "Investissements d'Avenir" Program (ANR 11 – LABX-0002-01).

# Introduction

The Julia Quick Syntax Reference (Second Edition) serves as a concise yet comprehensive guide to the Julia programming language, providing both new and experienced users with essential knowledge to leverage Julia's capabilities in data science, scientific computing, and general-purpose programming. Building on the 2019 first edition, this updated edition incorporates the latest developments in the Julia ecosystem, enhancing its relevance and usefulness for modern programming challenges.

This book is designed for programmers, data scientists, and researchers looking for a quick and accessible resource on Julia's syntax and features. Whether you are transitioning from languages such as Python, R, or MATLAB, or just beginning your programming journey, this guide provides a practical roadmap for mastering Julia's unique strengths.

This second edition corrects, updates, and expands on the first: it fixes typos from the first edition, updates the APIs of key packages in the ecosystem - many of which are now stable - and updates the discussion of tools for installing and using Julia (e.g. juliaup, VSCode). More importantly, the second edition expands on topics of growing importance.

The second edition delves deeper into modules and environments, introduces a comprehensive chapter on running Julia code with various forms of parallelism (GPU programming, multithreading, and multiprocessing), and thoroughly examines random number generation and reproducibility in stochastic computations.

It also includes a comprehensive introductory chapter on artificial intelligence, exploring its nature, scope and practical implementation of machine learning workflows in Julia using the BetaML package. This chapter focuses on building intuition for the algorithms underlying regression, classification, clustering, dimensionality reduction, missing data imputation, deep neural networks, and model interpretation.

INTRODUCTION

The book is divided into two main sections:
The first section is devoted to the core aspects of the language. Chapters 1-8 cover fundamental aspects of Julia, including installation, basic syntax, data types, control flow, user-defined types, and efficient coding techniques. Notable updates include expanded discussions of modules, environments, and the role of reproducibility in stochastic computation. Chapter 6 then introduces Julia's metaprogramming capabilities, a prominent feature that allows the manipulation of code as data, enabling dynamic and flexible programming paradigms.

Chapter 7 concludes this section with comprehensive guidance on integrating Julia with other languages such as Python, R, and C++.

The second section (chapters 9-13) explores Julia's extensive package ecosystem, introducing tools for data manipulation (DataFrames, IndexedTables), optimisation (JuMP), symbolic computation (SymPy), and machine learning (BetaML). A new chapter on artificial intelligence provides practical workflows for common ML tasks, including regression, classification, and neural networks.

The book strikes a balance between theory and practice, so that novice programmers can follow along, while experienced users will find advanced topics and insights to deepen their expertise. Whether your goal is scientific discovery, software development, or data-driven decision making, the Julia Quick Syntax Reference will guide you every step of the way.

# PART I

# Language Core

# CHAPTER 1

# Getting Started

In this chapter you will learn about the main features of the language and how to install Julia and the tools needed to develop programs in Julia on your computer. To emphasise their importance, the concepts of modules, packages and environments are already introduced in this first chapter.

## 1.1 Why Julia

With so many programming languages available, why create yet another one? Why invest the time to learn Julia? Is it worth it?

One of the main arguments in favor of using Julia is that it contributes to improving a trade-off that has long existed in programming: fast coding versus fast execution.

On the one side, Julia allows the developer to code in a dynamic, high-level language similar to Python, R, or MATLAB, interacting with the code and having powerful expressivity (see Chapter 6, for example). On the other side, with minimum effort, developers can write programs in Julia that run (almost) as fast as programs written in compiled languages such as C or FORTRAN, as you'll see in Chapter 8.

Wouldn't it be better, though, to optimize existing languages, with their large number of libraries and established ecosystems, rather than create a new language from scratch?

Well, yes and no. Attempts to improve runtime execution of dynamic languages are numerous: PyPy (https://pypy.org/), Cython (https://cython.org/), and Numba (https://numba.pydata.org/) are three notable examples for the Python programming language. They all clash with one fact: Python (and, in general, all the current dynamic languages) was designed before the recent development of just-in-time (JIT) compilers, and hence it offers features that are not easy to optimize. The optimization tools either fail or require complex workarounds in order to work.

Conversely, Julia has been designed from the ground up to work with JIT compilers, and the language features—and their internal implementations—have been carefully considered in order to provide the programmer with the expected productivity of a modern language, all while respecting the constraints of a compiler. The result is that Julia-compliant code is guaranteed to work with the underlying JIT compiler, producing in the end highly optimized compiled code. Indeed, if you check the languages in which the most popular numerical libraries in R or Python are actually coded (for example, by going to their GitHub repositories), you will notice that most have a "core" written in a compiled language, while the Julia packages are coded 100% in Julia. This is important because it helps to "lower the barrier" between the users of a numerical library and its developers.

---

 **Note**

**The Shadow Costs of Using a New Language**

If it is true that the main "costs" of using a new language relate to learning the language and having to abandon useful libraries and comfortable, feature-rich development editors that you are accustomed to, it is also true that in Julia these costs are mitigated by several factors:

- The language has been designed to syntactically resemble mainstream languages (you'll see it in this book!). If you already know a programming language, chances are you will be at ease with the Julia syntax.

- Julia allows you to easily interface your code with all the major programming languages (see Chapter 7), enabling you to reuse their huge sets of libraries (that is, when they are not already ported to Julia).

- The integrated development environments (IDEs) that are already available—for example, Visual Studio Code (VS Code) Julia extension and IJulia Jupyter kernel—are frankly quite cool, with many common features already implemented. They allow you to be productive in Julia from the first time you code with it.

---

Apart from the breakout in runtime performances from traditional high-level dynamic languages, the fact that Julia was created from scratch means it uses the best, most modern technologies, without concerns over maintaining compatibility with existing code or internal architectures. Some of the features of Julia that you will likely appreciate include a built-in Git-based package manager, full code introspection, multiple dispatches, in-core high-level methods for parallel computing, and Unicode characters in variable languages (e.g., Greek letters).

Thanks to its computational advantages, Julia has its natural roots in the domain of scientific, high-performance programming, but it is becoming more and more mature as a general-purpose programming language. This is why this book does not focus specifically on the mathematical domain, but instead develops a broad set of simple, elementary examples that are accessible even to beginner programmers.

## 1.2 Installing Julia

You can find the Julia version for your operating system in the Download section (https://julialang.org/downloads/) of the Julia Project website (https://julialang.org/). Depending on your OS, you are then instructed to download Juliaup (https://github.com/JuliaLang/juliaup), an installation and version manager that allows you to install and update multiple versions at the same time, using the Microsoft Store (https://www.microsoft.com/store/apps/9NJNWW8PVKMN) in Windows or curl in Linux or Mac.

> **❗ Important !**
>
> When installing Julia, make sure that you select the option to put Julia in your path (it isn't selected by default).

Juliaup (or a manual download of the official binaries, should you not wish to use the installation manager) will provide a Julia interpreter console (a.k.a., the "REPL"—Read, Eval, Print, Loop—terminal), where you can run Julia code in a command-line fashion.

For a better experience, check out an integrated development environment, such as the previously mentioned VS Code Julia extension, an IDE based on the popular VS Code (https://code.visualstudio.com/) editor, or IJulia (https://github.com/JuliaLang/IJulia.jl), the Julia Jupyter (http://jupyter.org/) back end. You can find detailed setup instructions on their respective websites, but in a nutshell the steps are pretty straightforward.

For VS Code:

1. Install the main Julia binaries (or Juliaup) first.
2. Download, install, and open the VS Code (https://code.visualstudio.com/) editor.
3. From within VS Code, click the **Extensions** icon in the Activity Bar and install the Julia extension.

For IJulia:

1. Install the main Julia binaries (or Juliaup).

2. Install the Python-based Jupyter Notebook server using your favorite OS-specific tool (e.g., a package manager in Linux, the Python pip package manager, or the Anaconda distribution).

3. From a Julia console, type the following:

   ```
   using Pkg; Pkg.update(); Pkg.add("IJulia");
   Pkg.build("IJulia")
   ```

The IJulia kernel is now installed. Just start the notebook server and access it using a browser. Note that, alternatively, you can skip installing the Jupyter server yourself. In that case IJulia will download and install a private-to-Julia Jupyter notebook that you can start by typing `using IJulia; notebook()` in a Julia session.

Another interesting option is to use Julia with Pluto (https://plutojl.org/). This is a reactive programming environment in which the notebooks are written as regular Julia files. I don't cover Pluto in this book, as it is beyond the scope of a quick syntax reference, but many data scientists appreciate its reactive nature, meaning that each cell is automatically updated as each other cell is evaluated, similar to a spreadsheet. This guarantees that the state of the whole notebook is always consistent, and if you change the value of one cell, you will see its effect across the whole notebook.

You can also choose, at least to start with, *not* to install Julia at all, and try instead one of the online computing environments that support Julia, such as JuliaHub (https://juliabox.com/), CoCalc (https://cocalc.com/doc/software-julia.html), Nextjournal (https://nextjournal.com/nextjournal/julia-environment), or Binder (https://mybinder.org/).

I wrote and tested the code in this book using Julia versions 1.10 and 1.11. The name and version of the various packages are declared at the top of each chapter. Furthermore, in the Git repository of the code that accompanies the book, two small files, `Project.toml` and `Manifest.toml`, are stored for each chapter folder. This will allow you to use the exact same environment that I used to write the code in the book (environments will be discussed further in this chapter).

## 1.3 Running Julia

There are many alternative ways to run Julia code, depending on your needs:

1. Run Julia interactively in a console. Start `julia` in a terminal to obtain the REPL console, and then type the commands there (type `exit()` or press Ctrl+D when you have finished).

2. Create a script—a text file ending in `.jl`—and let Julia parse and run it with `julia myscript.jl [arg1, arg2,..]`. You can also run script files from within the Julia console by typing `include("myscript.jl")`.

3. In Linux or macOS, you could instead add at the top of the script the location of the Julia interpreter on your system, preceded by `#!` and followed by an empty row; for example, `#!/usr/bin/julia` (you can find the full path of the Julia interpreter by typing `which julia` in a console). Be sure that the file is executable (e.g., run `chmod +x myscript.jl`). You can then run the script with `./myscript.jl`.

4. Use an IDE (such as those previously mentioned), open a Julia script, and use the run command specific to the editor.

You can define both a global (for all users of the computer) and a local (for a single user) Julia file that will be executed at any startup, which enables you to, for example, define functions or variables that should be always available. The following are the locations of these two files:

- **Global Julia startup file**: [JULIA_INSTALL_FOLDER]\etc\julia\startup.jl (where JULIA_INSTALL_FOLDER is the folder in which Julia is installed)
- **Local Julia startup file**: [USER_HOME_FOLDER]\.julia\config\startup.jl (where USER_HOME_FOLDER is the home folder of the local user; e.g., %HOMEPATH% in Windows and ~ in Linux)

Remember to use the path with forward slashes (/) with Linux. Note that the local config folder may not exist. In that case, just create the config folder as a .julia subfolder and start the new startup.jl file there.

---

 **Tip**

Julia keeps all the objects created within the same work session in memory. You may sometimes want to free memory or "clean up" your session by deleting no longer needed objects. If you want to do this, just restart the Julia session (you may want to use the trick mentioned at the end of Chapter 3) or use the Revise.jl (https://github.com/timholy/Revise.jl) package for finer control.

---

Within a Julia session, you can check which version of Julia you are using with versioninfo().

CHAPTER 1   GETTING STARTED

## 1.4  Miscellaneous Syntax Elements

Julia supports both single-line (#) and multiline (#= [...] =#) comments. Multiline comments can be nested and appear anywhere in the line:

```
println("Some code..")
#=
   Multiline comment
   #= nested multiline comment =#
   Still a comment
=#
println(#= A comment in the middle of the line =# "This is a code") # Normal single-line comment
```

You don't need to use semicolons to indicate the end of a statement. If they're used, semicolons will suppress the command output (this is done automatically in scripting mode). If the semicolon is used alone in the REPL, it allows you to switch to the OS command shell prompt in order to launch a system-wide command.

Blocks don't need to be enclosed in brackets. However, they do require the keyword end.

---

> **❗ Important !**
>
> While indentation doesn't carry any functional meaning in the Julia language, empty spaces sometimes are important. For example, function calls must have the opening parenthesis with the inputs strictly attached to the function name, such as in the second example here:

```
println (x)  # rise an ERROR
println(x)   # OK
```

In Julia, variable names can include a wide subset of Unicode symbols, allowing a variable to be represented, for example, by a Greek letter.

In most Julia development environments (including the REPL), to type a Greek letter you can use a LaTeX-like syntax. This involves typing \, then the LaTeX name for the symbol (e.g., \alpha for $\alpha$), and finally pressing the Tab key to confirm. Using LaTeX syntax, you can also add subscripts, superscripts, or decorators.

All the followings are valid variable names: $x_1$, $\tilde{x}$, $\alpha$, y1, $y^{(a+b)}$, y2. Note, however, that although you can use y2 as a variable name, you can't use 2y, as the latter is automatically interpreted as 2 * y. Together with Unicode, this greatly simplifies the transposition in computer code of mathematical equations. While Unicode symbols can improve readability in your own script, it is not recommended to write APIs using them. That's because if you want to use your Julia code in other languages that don't support Unicode symbols, you'll have to wrap the API using other names.

If you come from a language that follows a zero-indexing standard (e.g., C or Python), one important thing to remember is that Julia arrays are one-based indexed (counting starts from 1 and not 0). There are ways to override this behavior, but in many cases doing so probably would do more harm than good.

## 1.5 Modules and Packages

Julia developers have chosen an approach that makes the core of Julia relatively light, and additional functionality is usually provided by external packages.

Julia binaries ship with a set of these packages (think of them as a "Standard Library") and a powerful package manager that can download (typically directly from GitHub repositories), pre-compile, update, and solve dependencies, all with a few simple commands.

While *registered* packages can be installed simply by using their name, *unregistered* packages need their source location to be specified. At the time of this writing, over 10,000 registered packages have been published.

Knowing how packages work is essential to efficiently working in Julia, and this is why I have chosen to introduce package management early in the book and complement the book with a discussion of some common packages.

## 1.5.1 Using the Package Manager

There are two ways to access package management features, interactively and programmatically from within other Julia code:

- **Interactively**: Type ] in the REPL console to enter a "special" pkg mode. The prompt then changes from julia> to ([CURRENT_ENVIRONMENT]) pkg>, where [CURRENT_ENVIRONMENT] (see section 1.6 later in this chapter) defaults to vX.Y, the current version of Julia. You can then run package manager commands at the prompt or go back to the normal interpreter mode by pressing Backspace.

- **Programmatically**: Import the Pkg module into your code (with using Pkg) and then run Pkg.command(ARGS). Obviously, nothing inhibits you from using the programmatic way in an interactive session, but the special package mode has tab completion and other goodies that make it more comfortable to use.

Also note that the two interfaces are not 100% consistent, with the programmatic interface being slightly more stringent.

Some of the useful package commands are explained in the following list:

- `status`: Retrieves a list (name and version) of the locally installed packages.
- `update`: Updates the local index of packages and all the local packages to the latest version.
- `add pkgName`: Automatically downloads and installs a given package. For multiple packages, use `add Pkg1 Pkg2` or `Pkg.add(["Pkg1","Pkg2"])`.
- `add pkgName#main`: Checks out the main branch of a given package.
- `add pkgName#branchName`: Checks out a specific branch of a given package.
- `add pkgName#vX.Y.Z`: Checks out a specific release of a given package.
- `free pkgName`: Returns the package to the latest release.
- `rm pkgName`: Removes a package and all its dependent packages that have been installed automatically only for it.
- `add git@github.com:userName/pkgName.jl.git`: Checks out a nonregistered package from a Git repository (here, it's GitHub).

## 1.5.2 Using Modules

Before we delve into using packages, this section introduces modules, because in many ways packages are just modules with some metadata attached to them that allows the module to be easily referenced from other people's code.

*Modules*, like functions, allow code that performs some functionality to be isolated so that it can be written once and used in multiple places. Perhaps more importantly, modules allow you to partition the variable name space into different *namespaces*, so that you can have, for example, `Module1.apple` that refers to a different object than `Module2.apple`.

Modules can contain variables, functions, constants, other (sub) modules (rarely used in Julia), and so forth, and are introduced with the keyword `module`. They are usually written in CamelCase, and the module content is not indented. The following is an example for purposes of explanation:

```
module Foo
export plus2, an_exported_var
an_exported_var = 2
a_private_var = 5
plus10(x) = x + a_private_var + 10
plus2(x) = x + a_private_var + 2
println("I am in Foo, I can access $a_private_var")
end
```

This module defines two variables and two functions, plus a call to the base function `println`. For now, ignore how we defined the functions and the interpolation, which we'll explore in detail in the subsequent chapters. Here we will focus on how to access the objects (variables and functions) provided by the `Foo` module.

First, you need to *evaluate* the code corresponding to the Foo module. You can copy and paste it into the REPL or select it in VS Code and run it by pressing Shift+Enter. Another way is to save it in a file like Foo.jl and use the include("Foo.jl") command. At this point the code is parsed and evaluated, as you can see from the previous output of the print call.

---

### ⓘ Note

You can also use include in the module Foo; that is, there is no need to keep one module equal to one file. You can split a module into as many files as you like, or use one file to host several modules. However, because include evaluates the included code, make sure that your code is included only once in your workflow; otherwise, you may end up with multiple copies of the same module. If you need to make changes to the included code during module or package development, use includet from the Revise.jl (https://github.com/timholy/Revise.jl) package instead, which will automatically track the changing code.

---

You can also see in the output that the Foo module is referred to as Main.Foo. Indeed, the "root" module, the one you are in when you start the REPL or VS Code, is called Main, and now Foo is a child of it. You can refer to its elements either by their full paths, (e.g., Main.Foo.a_private_var) or by their relative paths (e.g., Foo.a_private_var). Alternatively, you can inject the exported names directly into the current namespace with using Main.Foo or using .Foo (note the dot!) and then access them directly, such as by using an_exported_var.

CHAPTER 1   GETTING STARTED

## 1.5.3 Using Packages

Using a package is similar to using a module that you have evaluated yourself. The difference is that typing using PackageName (without the dot) does a couple of things before injecting the exported names of PackageName into the current namespace: first it looks up the module PackageName from a known location (where it was placed when originally added with the package manager), and then it evaluates it.

Alternatively, import PackageName does the lookup and evaluation, but not the name injection, which helps to keep the namespace clean. You can also use aliases, or choose to import only a subset of functions (which you can then access directly).

For example, to access the function plot() available in the package Plots.jl, you can do any of the following (see the "Plotting" section in Chapter 9 for specific plotting issues):

- Access the package function directly with using PackageName:

```
using Plots
plot(rand(4,4))
```

- Access the package functions using their full name with import PackageName:

```
import Plots as pl
pl.plot(rand(4,4)) # `Equivalent to Plots.
                        plot(rand(4,4))`
```

Note that the aliasing part in import Plots as pl requires Julia >= v1.6.

- Access the package functions directly with `import PackageName:package_function`:

```
import Plots: plot # You can import multiple
                     functions at once using commas

plot(rand(4,4))
```

You can read more about packages in the relevant section (https://julialang.github.io/Pkg.jl/v1/) of the Julia documentation, or just typing `help` or `help COMMAND` in pkg mode to get more details on package manager commands.

---

> 🛑 **Important !**
>
> Across this book I will refer to several packages, including packages in the Julia Standard Library and third-party packages. When I state that a given function belongs to a given package, remember to add `using PackageName` to run the code in the examples (I will not repeat this instruction every time).

---

## 1.6 Environments

Julia has a very lightweight but effective implementation of the "environment" concept. Packages (possibly in multiple versions) are "physically" stored centrally in a user-defined folder. Indeed, when you create an environment, only two small text files, `Project.toml` and `Manifest.toml`, are created. These files list, respectively, the packages used directly in the project (and possibly the version ranges known to work with the project) and the list and exact version of each package used.

CHAPTER 1   GETTING STARTED

In practice, if you have tens, hundreds, or even thousands of projects that you work with on your computer, you always create a specific environment for each of these projects, because it is very "cheap" (the packages are not copied, as is the case with other languages). If you then give the end user (or yourself, if returning to the project much later) your script files plus these two small files, the user will be able, with a simple command, to be back in exactly the same environment that you were in when you were developing and running your Julia project. This is awesome for reproducibility!

How exactly do you use environments? You start by activating a specific directory (often the current directory) with ] activate . (where the ] prompt indicates that you have entered the package mode). To revert to the default shared environment, type ] activate with no arguments.

At this point you can add a package to this environment with add PackageName. For example, if you start a new environment and add the CSV.jl and DataFrames.jl packages, your Project.toml file would look like this:

```
----------------------------------------------------------------
[deps]
CSV = "336ed68f-0bac-5ca0-87d4-7b16caf5d00b"
DataFrames = "a93c6f00-e57d-5684-b7b6-d8193f3e46c0"
----------------------------------------------------------------
```

And your Manifest.toml file would look like this:

```
----------------------------------------------------------------
# This file is machine-generated - editing it directly is
not advised

julia_version = "1.11.0"
manifest_format = "2.0"
project_hash = "7e38425d15a28e7abd87534dcfc793c08d63a4f4"
```

```
[[deps.Artifacts]]
uuid = "56f22d72-fd6d-98f1-02f0-08ddc0907c33"

[[deps.Base64]]
uuid = "2a0f44e3-6c83-55bd-87e4-b1978d98bd5f"

[[deps.CSV]]
deps = ["CodecZlib", "Dates", "FilePathsBase", "InlineStrings",
"Mmap", "Parsers", "PooledArrays", "PrecompileTools",
"SentinelArrays", "Tables", "Unicode", "WeakRefStrings",
"WorkerUtilities"]
git-tree-sha1 = "679e69c611fff422038e9e21e270c4197d49d918"
uuid = "336ed68f-0bac-5ca0-87d4-7b16caf5d00b"
version = "0.10.12"

...etc..
```
---

As you can, whereas Project.toml contains only the name (and ID) of the packages you have installed, Manifest.toml contains all the details of their dependencies.

If you get a project from someone and want to "re-create" that environment, all you have to do is activate it and use ] instantiate to get all the necessary packages downloaded and installed (if needed).

Unfortunately, the reproducibility guarantee only works within the same version of Julia. If you activate an environment from a different version of Julia, you should first run ] resolve to automatically find the combination of package versions that work for your version of Julia. As of Julia 1.11, you can have a specific manifest file per Julia version by simply renaming Manifest.toml to Manifest-vX.Y.toml.

# CHAPTER 1 GETTING STARTED

In practice, I start all my projects with the following code:

---
```
cd(@__DIR__) # set the current directory to the directory path
             of the file
using Pkg
Pkg.activate(".")
#Pkg.resolve()
Pkg.instantiate()

# Rest of my script...
```
---

 **Warning !**
**Current Directory vs. Environment**

Do not confuse the *current directory* (`cd("dir")`) with the *environment directory* (`Pkg.activate("dir")`). While they are often the same, the current directory defines the relative path when you are working on the file system, such as for opening or saving a file. The environment directory specifies instead where the `Project.toml` and `Manifest.toml` files that define the environment are located.

In the GitHub repository that accompanies this book, I provide the `Project.toml` and `Manifest.toml` files so that you can re-create the environment I used while writing the book and get all the examples working.

## 1.7 Help System

Julia comes with an integrated help system that retrieves usage information for most functions directly from the source code. This is true also for most third-party packages.

Typing ? in the console leads to the Julia help system, and the prompt changes to help?>. From there you can search for the function's API.

---

 **Tip**

In noninteractive environments like IJulia notebooks, you can use ?search_term to access the documentation.

If you don't remember exactly the function name, Julia is kind enough to return a list of similar functions. In VS Code, just hover your mouse cursor over the code to quickly access the available documentation, or right-click the code and select **Show Documentation** to get it printed in the documentation panel.

---

While the actual content returned may vary, you can expect to have the following information for each function you query:

- Its signature
- One-line description
- Argument list
- Hints to similar or related functions
- One or more usage examples
- A list of methods (for functions that have multiple implementations)

# CHAPTER 2

# Data Types and Structures

The following third-party package is covered in this chapter:

StableRNGs.jl  https://github.com/JuliaRandom/StableRNGs.jl   v1.0.2

Julia natively provides a fairly complete and hierarchically organized set of predefined types (especially the numerical ones). These are either *scalar*, such as integers, floating-point numbers, and chars, or *container-like* structures that are able to hold other objects, such as multidimensional arrays, dictionaries, sets, and so on.

In this chapter we discuss the predefined types, while in Chapter 4, where we cover Julia custom types, we consider the hierarchical organization of both predefined types and custom types.

Every value (even primitive ones) has its own unique type. By convention, types start with a capital letter, such as `Int64` or `Bool`. Sometimes (such as for all container-like structures and some non-container structures), the name of the type is followed by other *parameters* inside curly brackets, like the types of the contained elements or the number of dimensions. For example, `Array{Int64,2}` would be used for a two-dimensional array of integers.

In Julia terminology these are referred as to *parametric types*. In this book we will use T as a placeholder to generically indicate a type.

There is no division between object values and non-object values. All values in Julia are true objects having a type. Only values, not variables, have types. Variables are simply names bound to values. The :: operator can be used to attach type annotations to expressions and variables in programs. There are two primary reasons to do this:

- As an assertion to help confirm that your program works the way you expect

- To provide extra type information to the compiler, which can then, in some limited cases, improve performance

## 2.1 Simple Types (Non-Containers)

Individual characters are represented by the Char type, such as a = 'a' (Unicode is fully supported). Boolean values are represented by the Bool type, whose unique instances are true and false.

> **❗ Important !**
>
> In Julia, single and double quotes are not interchangeable. Single quotes produce a Char (e.g., x = 'a'), whereas double quotes produce a String (e.g., x = "a").
>
> While the Boolean values true and false in an integer context are automatically cast and evaluated as 1 and 0, respectively, the opposite is not true: if 0 [...] end would raise a "non-boolean (Int64) used in boolean context" TypeError.

CHAPTER 2   DATA TYPES AND STRUCTURES

The "default" integer type in Julia is `Int64` (there are actually ten different variants of integer types), and it is able to store values between $-2^{63}$ and $2^{63-1}$ (to identify the minimum and maximum values possible for a given type, use `typemin(T)` and `typemax(T)`; e.g., `typemax(Int64)`). Similarly, the "default" floating-point type is `Float64`. Complex numbers (`Complex{T}`) are supported through the global constant `im`, representing the principal square root of -1. A complex number can then be defined as `a = 1 + 2im`. The mathematical focus of Julia is evident by the fact that there exists a native type even for exact ratios of integers, `Rational{Int64}`, whose instances can be constructed using the `//` operator: `a = 2 // 3`.

## 2.1.1  Basic Mathematical Operations

All standard basic mathematical arithmetic operators are supported in the obvious way (`+`, `-`, `*`, `/`). To raise a number to a power, use the `^` operator (e.g., `a = 3^2`). Natural exponential expressions (i.e., with the Euler's number as a base) are created with `a = exp(b)` or using the global constant $e$ (this is *not* the normal letter e, but the special Unicode symbol for the Euler's number: type `\euler` and press the Tab key in the REPL to obtain it or use `MathConstants.e`). Integer divisions are implemented with the ÷ operator (`\div` + Tab) and their remainders are given using the "modulo" `%` operator (e.g., `a = 3 % 2`). The pi constant is available as a global constant `pi` or π (`\pi` + Tab).

## 2.2  Strings

The `String` type in Julia can be seen in some ways as a specialized array of individual chars (for example, strings support indexing or looping over their individual letters like an `Array` would). Unlike arrays, strings are immutable (e.g., `a="abc"; a[2]='B'` would raise an error).

## CHAPTER 2  DATA TYPES AND STRUCTURES

A string on a single row can be created using a single pair of double quotes, while a string on multiple rows can use a triple pair of double quotes:

```
a = "a string"
b = "a string\non multiple rows\n"
c = """
a string
on multiple rows
"""
a[3] # Returns 's'
```

Julia supports most typical string operations. For example:

- `split(s, " ")` defaults to whitespace.
- `join([s1,s2], "")`
- `replace(s, "toSearch" => "toReplace")`
- `strip(s)` removes leading and trailing whitespace.
- `lowercase(s)` / `uppercase(a)`

To convert strings representing numbers to integers or floats, use `myInt = parse(Int,"2017")`. To do the opposite and convert integers or floats to strings, use `myString = string(123)`.

## 2.2.1 Concatenation

There are several ways to concatenate strings:

- The concatenation operator: *
- The string function: `string(str1,str2,str3)`

- Using *interpolation*; that is, combining string variables using the dollar sign: `a = "$str1 is a string and $(my_object.int1) is an integer"` (note the use of parentheses for larger expressions)

> **❗ Important !**
>
> While both the `string` function and the use of interpolation automatically cast compatible types (e.g., `Float64` and `Int64`) into strings, the concatenation operator `*` doesn't. Be careful not to mistake the `string` function (with a lowercase s) with the `String` type and the homonymous constructor (with a capital S).

## 2.3 Arrays (Lists)

Arrays (`Array{T,nDims}`) are n-dimensional mutable containers. In this section, we deal with one-dimensional arrays (e.g., `Array{Float64,1}`). In the next section we consider two-dimensional (or more) arrays (e.g., `Array{Float64,2}`).

> **💡 Tip**
>
> In Julia, you may also encounter *vectors* and *matrices*. `Vector{T}` is just an alias for a one-dimensional `Array{T,1}`, and `Matrix{T}` is an alias for a two-dimensional `Array{T,2}`.

## CHAPTER 2  DATA TYPES AND STRUCTURES

There are several ways to create an Array of type T:

- **Column vector (one-dimensional array):**
  a = [1;2;3] or a = [1,2,3]
- **Row vector:** a = [1 2 3] (in Julia, this is a two-dimensional array where the first dimension is made of a single row, as described in the next section, "Multidimensional and Nested Arrays")
- **Empty (zero-element) arrays:**
  - a = [] (resulting in Array{Any,1})
  - a = T[] (e.g., a = Int64[])
  - a = Array{T,1}() (using the constructor explicitly)
  - c = Vector{T}() (using the Vector alias)
- *n*-**elements zeros array:** a = zeros(n) or a = zeros(Int64,n)
- *n*-**elements ones array:** a = ones(n) or a = ones(Int64,n)
- *n*-**elements array whose content is garbage:**
  a = Array{T,1}(undef,n)
- *n*-**elements array of identical j elements:**
  a=fill(j, n)
- *n*-**elements array of random numbers:** a = rand(n)

Arrays are allowed to store heterogeneous types, but in such cases, the array will be of type Any and it will be in general much slower; for example, x = [10, "foo", false]. However, if you need to store a limited set of types in the array, you can use the Union keyword and still have an efficient implementation; for example., a = Union{Int64,String,Bool}

CHAPTER 2   DATA TYPES AND STRUCTURES

[10, "Foo", false]. a = Int64[] is just a shorthand for a = Array{Int64,1}() (e.g., a = Any[1,1.5,2.5] is equivalent to a = Array{Any,1}([1,1.5,2.5])).

---

> **! Caution**
>
> a = Array{Int64,1} (without the parentheses) doesn't create an Array at all, but just assigns the DataType Array{Int64,1} to a.

---

Square brackets are used to access the elements of an array (e.g., a[1]). The slice syntax from:step:to is generally supported and in several contexts will return a (fast) iterator rather than an array. To then transform the iterator into an array, use collect(myIterator), such as a = collect(5:10). Bounding limits are included in the sequence. For example, collect(4:2:8) will return the array [4,6,8] and collect(8:-2:4) will return [8,6,4]. Together with the keyword end, you can use [...] the slice operator (the colon punctuaction mark :) to reverse [...] an array (although you are probably better off with the dedicated function reverse(a)): collect(a[end:-1:1]).

You can initialize an array with a mix of values and ranges with either y=[2015; 2025:2030; 2100] (note the semicolons!) or the vcat command (which stands for *vertical concatenation*): y=vcat(2015, 2025:2030, 2100).

The following functions are useful while working with arrays:

---

>  **Tip**
>
> By convention, functions that end with an exclamation mark will modify the first of their arguments.

---

- push!(a,b): Push an element to the end of a. If b is an Array, it will be added as a single element to a (equivalent to Python append).
- append!(a,b): Append all the elements of b to a. If b is a scalar, push! and append! are interchangeable. Note that a string is treated as an array by append!. It is equivalent to Python extend or +=.
- c = vcat(1,[2,3],[4,5]): Concatenate arrays.
- pop!(a): Remove an element from the end.
- popfirst!(a): Remove an element at the beginning (left) of the array.
- deleteat!(a, pos): Remove an element at an arbitrary position.
- pushfirst!(a,b): Add b at the beginning (left) of a (and no, sorry, appendfirst! doesn't exist).
- sort!(a, rev=false) or sort(a, rev=false): Sort a (depending on whether you want to modify the original array).
- sortperm(a, rev=false): Return the index of the sorted elements (without actually sorting them).
- unique!(a) or unique(a): Remove duplicates (depending on whether you want to modify the original array).
- reverse(a) or a[end:-1:1]: Reverse the order of elements in the array.
- in(b, a): Check for the existence of b in a. Also available as an operator: if b in a [...] end.

CHAPTER 2  DATA TYPES AND STRUCTURES

- length(a): Get the length of a.

- a... (the "splat" operator): Convert the values of an array in function parameters (used inside function calls; see the next item).

- maximum(a) or max(a...): Get the maximum value (max returns the maximum value between the given arguments).

- minimum(a) or min(a...): Get the minimum value (min returns the minimum value between the given arguments).

- sum(a): Return the summation of the elements of a.

- cumsum(a): Return the cumulative sum of each element of a (i.e., it returns an Array).

- empty!(a): Empty an array (works only for column vectors, not for row vectors).

- b = vec(a): Transform row vectors into column vectors.

- shuffle(a) or shuffle!(a): Randomly shuffle the elements of a (requires using Random).

- isempty(a): Check if an array is empty.

- findall(x -> x == value, myarray): Find a value in an array and return its indexes. This is a bit tricky. The first argument is an anonymous function (see Chapter 3) that returns a boolean value for each value of myarray, and then findall() returns the index positions.

31

- `deleteat!(myarray, findall(x -> x == my_unwanted_item, myarray))`: Delete a given item from a list.
- `enumerate(a)`: Get *(index,element)* pairs; that is, return an iterator to tuples, where the first element is the index of each element of the array a and the second is the element itself (used, for example, in `for (idx,v) in enumerate(something) ... end` loops).
- `zip(a,b)`: Get *(a_element, b_element)* pairs; that is, return an iterator to tuples made of elements from each of the arguments (e.g., `zip(names,gender,age)` would result in something like `[("Marc",'M',18),("Anne",'F',16)]`). Note that, like enumerate, an *iterator* is returned.

## 2.3.1 Multidimensional and Nested Arrays

In this section we deal with multidimensional arrays (e.g., two-dimensional array `Array{T,2}` or `Matrix{T}`) and nested arrays, as an array of arrays `Array{Array{T,1},1}` (i.e., the main structure remains a one-dimensional array, but the individual elements are themselves arrays). The main difference between a *matrix generic* and an *array of arrays* is that, with a matrix, the number of elements on each column (row) must be the same and rules of linear algebra apply. As in matrix algebra, the first dimension is interpreted as the vertical dimension (rows) and the second dimension is the horizontal one (columns).

Multidimensional arrays `Array{T,N}` can be created similarly to one-dimensional arrays (indeed, the latter are just specific cases of the former):

- **By column**: a = [[1,2,3] [4,5,6]] (elements of the first column, elements of the second column, and so on). Note that this is valid only if you write the matrix in a single line.

- **By row**: a = [1 4; 2 5; 3 6] (elements of the first row, elements of the second row, and so on).

- **Generic n-dimensional arrays**: Can be hard-coded by specifying ; as the separator of the first dimension (rows), ;; as the separator of the second dimension, and so on. For example, [1;2;3;; 10;20;30;;; 4;5;6;; 40;50;60;;;] creates a 3×2×2 array whose first column of the first "face" is [1,2,3] (requires Julia >= v1.7).

- **Empty (zero-elements) arrays**: a = Array{T}(undef, 0, 0, 0).

- ***n,m,g*-elements zeros array**: a = zeros(n,m,g) or a = zeros(Int64,n,m,g).

- ***n,m,g*-elements ones array**: a = ones(n,m,g) or a = ones(Int64,n,m,g).

- ***n,m,g*-elements array whose content is garbage**: a = Array{T,3}(undef,n,m,g).

- ***n,m,g*-elements array of identical *j* elements**: a = fill(j,n,m,n).

- ***n,m,g*-elements array of random numbers**: a = rand(n,m,g).

Multidimensional arrays often arise from using list comprehension. For example, a = [3x + 2y + z for x in 1:2, y in 2:3, z in 1:2] (see Chapter 3 for details on list comprehension).

CHAPTER 2   DATA TYPES AND STRUCTURES

> **❗ Important !**
> **Matrix vs. Nested Arrays**
>
> Note this important difference:
>
> - a = [[1,2,3],[10,20,30]] creates a one-dimensional array with two elements (each of those is again a vector).
> - a = [[1,2,3] [10,20,30]] creates a two-dimensional array (a matrix with two columns) with six scalar values.

Nested arrays can be accessed with double square brackets, such as a[2][3].

Elements of n-dimensional arrays can be accessed instead with the a[idxDim1,idxDim2,..,idxDimN] syntax (e.g., a[row,col] for a matrix), where again the slice syntax can be used. For example, given that a is a 3×3 matrix, a[1:2,:] would return a 2×3 matrix with all the column elements of the first and second row. To select based on a specific dimension, use selectdim(array,dimension,index(es)). For example, selectdim(a,2,[1,2]) selects the first two columns of a.

Note that in order to push elements to an array, they have to be compatible with the array type. The following code won't work:

a = [1,2,3]; push!(a,1.5)

The reason it won't work is that the first command automatically creates an Array{Int64,1} (i.e., an array of integers), and then the second command tries to push on this array a float number (Julia tries to automatically cast 1.5 to an integer, but it fails with an InexactError). In order to work, a must be defined either as Float64[1,2,3] or as Union{Int64,Float64}[1,2,3].

34

CHAPTER 2  DATA TYPES AND STRUCTURES

Boolean selection is implemented using a boolean array (possibly multidimensional) for the selection:

---
```
a = [[1,2,3] [10,20,30]]
mask = [[true,true,false] [false,true,false]]
```
---

a[mask] returns a one-dimensional array with 1, 2, and 20. Note that boolean selection always results in a flattened array, even if you delete a whole row or a whole column of the original data. It is up to the programmer to reshape the data accordingly.

### ⓘ Note

For row vectors, both a[2] and a[1,2] return the second element.

Several functions are particularly useful when working with n-dimensional arrays (on top of those described in the previous section). They often include a dims keyword to specify on which dimensions they operate (recall that 1 is the row dimension):

- size(a): Return a tuple (i.e., an immutable list) with the sizes of the *n* dimensions.

- ndims(a): Return the number of dimensions of the array (e.g., 2 for a matrix).

- reshape(a, nElementsDim1, nElementsDim2,...,nElemensDimN): Reshape the elements of a in a new n-dimensional array with the dimensions given. Returns a DimensionMismatch error if the new dimensions are not consistent with the number of elements held by a.

35

- `dropdims(a, dims=(dimToDrop1,dimToDrop2))`: Remove the specified dimensions, provided that the specified dimensions have only a single element; for example, `a = rand(2,1,3); dropdims(a,dims=(2))`.
- `transpose(a)` or `a'`: Transpose a one- or two-dimensional numerical array (use `permutedims` for nonnumerical matrices).
- `sortslices(a,dims=1,by=x->(x[2],x[1]),rev=true)`: Sort (in reverse order) the rows of a matrix by first the second column and then the first one.

---

### ❗ Important !

Do not confuse `reshape(a)` with `transpose(a)`:

```
julia> a = [[1 2 3]; [4 5 6]]
2×3 Array{Int64,2}:
 1  2  3
 4  5  6

julia> reshape(a,3,2)
3×2 Array{Int64,2}:
 1  5
 4  3
 2  6

julia> transpose(a)
3×2 LinearAlgebra.Adjoint{Int64,Array{Int64,2}}:
 1  4
 2  5
 3  6
```

CHAPTER 2    DATA TYPES AND STRUCTURES

- collect(Iterators.flatten(a)), vec(a) or a[:]:
  Flatten a multidimensional array into a column vector.

> **Note**
>
> Note that reshape(), transpose(), and vec() perform a shadow copy, returning just a different "view" of the underlying data (so modifying the original matrix also modifies the reshaped, transposed, or flattened matrix). You can use collect() to force a *deep copy*, where the new matrix holds independent data (see the section "Memory and Copy Issues" later in this chapter for details).

- hcat(col1, col2): Concatenate horizontally anything that makes sense to be concatenated horizontally (vectors, matrices, data frames, etc.).
- hcat(col1, col2): Concatenate vertically anything that makes sense to be concatenated vertically (vectors, matrices, data frames, etc.).
- cat(array1, array2, dims=n): Generalize hcat and vcat to any arbitrary dimension (array1 and array2 must have the same size on all dimensions except possibly n).

## 2.4 Tuples

Use tuples (Tuple{T1,T2,...}) to create an immutable list of elements: t = (1,2.5,"a"). You can even do so without parentheses: t = 1,2.5,"a". *Immutable* refers here to the fact that once created, elements of this data structure cannot be added, removed, or modified (rebound to other objects). Unlike arrays, tuples may be stored in the stack part of the memory, and using them could be much more efficient.

> **💡 Tip**
>
> Note that if the element in a tuple is itself mutable (e.g., an `Array`), this is still allowed to mutate (i.e., "internally" change). What is "immutable" is the memory addresses or the actual bits of the various objects in the tuple.

Tuples can be easily unpacked to multiple variables: `var1, var2 = (x,y)`. This is useful, for example, for collecting the values of functions returning multiple values, where the returned object would be a tuple.

> **ⓘ Note**
>
> In stark contrast to arrays, tuples remain efficient even when hosting heterogeneous types, as the information on the types of each element hosted is retained in the type signature. For example, `typeof((1,2.5,"a"))` is not `Tuple{Any}` (as it would be for an array) but `Tuple{Int64,Float64,String}` (you will see why this matters in Chapter 8).

The following are some useful tricks:

- **Convert a tuple into an array**: `t=(1,2,3); a = [t...]` or `a = [i[1] for i in t]` or `a=collect(t)`
- **Convert an array into a tuple**: `t = (a...,)` (note the comma after the splat operator)

Note, however, that tuples cannot be used in linear algebra computations. If you need what, in many respects, are tuples, but with an array-like interface, you can use the package `StaticArrays.jl` (https://github.com/JuliaArrays/StaticArrays.jl).

## 2.5 Named Tuples

Named tuples (`NamedTuple`) are collections of items whose position in the collection (index) can be identified not only by their position but also by their name (the key):

- `nt = (a=1, b=2.5)`: Define a `NamedTuple`
- `nt.a`: Access the elements with the dot notation
- `keys(nt)`: Return a tuple of the keys
- `values(nt)`: Return a tuple of the values
- `collect(nt)`: Return an array of the values
- `pairs(nt)`: Return an iterable of the pairs (key/value); useful for looping: `for (k,v) in pairs(nt) [...] end`

As with "normal" tuples, named tuples can hold any values, but cannot be modified (i.e., they are "immutable"). They too remain efficient even when hosting heterogeneous types.

## 2.6 Dictionaries

Dictionaries (`Dict{Tkey,Tvalue}`) store mappings from keys to values and they have an apparently random sorting.

You can create an empty (zero-element) dictionary with `mydict = Dict()` or specify the key and values types with `Dict{String,Int64}()`. To initialize a dictionary with some values, use

`mydict = Dict('a'=>1, 'b'=>2, 'c'=>3)`

CHAPTER 2  DATA TYPES AND STRUCTURES

Here are some useful methods for working with dictionaries:

- `mydict[akey] = avalue`: Add pairs to the dictionary.
- `delete!(mydict,akey)`: Delete the pair with the specified key from the dictionary.
- `map((i,j) -> mydict[i]=j, ['a','b','c'], [1,2,3])`: Add pairs using maps (i.e., from vectors of keys and values to the dictionary).
- `mydict['a']`: Retrieve a value using the key (it raises an error if the looked-up key doesn't exist).
- `get(mydict,'a',0)`: Retrieve a value with a default value for a nonexistent key.
- `keys(mydict)`: Return all the keys; the result is an iterator, not an array. Use `collect()` to transform it.
- `values(mydict)`: Return all the values (result is again an iterator).
- `haskey(mydict, 'a')`: Check if a key exists.
- `in(('a' => 1), mydict)`: Check if a given key/value pair exists (that is, if the key exists and has that specific value).

You can iterate through both the key and the values of a dictionary at the same time:

```
for (k,v) in mydict
    println("$k is $v")
end
```

You can use this, for example, to look up the keys of a dictionary based on some value; for example:

`mykeys = [k for (k,v) in mydict if v==2]`

While named tuples and dictionaries can look similar, there are some important differences between them:

- Name tuples are immutable, while dictionaries are mutable.
- Dictionaries are type-unstable if different types of values are stored, while named tuples remain type-stable:
  - `d = Dict(:k1=>"v1", :k2=>2)` # Dict{Symbol,Any}
  - `nt = (k1="v1", k2=2)` # NamedTuple{(:k1, :k2),Tuple{String,Int64}}
- The syntax is a bit less verbose and readable with named tuples: `nt.k1` vs `d[:k1]`.

Overall, named tuple are generally more efficient and should be thought more as anonymous `struct` (see Chapter 4) than dictionaries.

## 2.7 Sets

Use sets (`Set{T}`) to represent collections of unordered, unique values. Sets are mutable collections.

Here are some methods:

- `s = Set()` or `s = Set{T}()`: Create an empty (zero-element) set.

- `s = Set([1,2,2,3,4])`: Initialize a set with values. Note that the set will then have only one instance of 2.
- `push!(s, 5)`: Add elements.
- `delete!(s,1)`: Delete elements.
- `intersect(set1,set2)`, `union(set1,set2)`, `setdiff(set1,set2)`: Set common operations of intersection, union, and difference, respectively.

## 2.8 Dates and Times

Julia has two data structures for representing dates and times, `Date` and `DateTime` (plus several subtypes of `Period`), both implemented in the Dates module of the Standard Library.

---

### Note

While a `DateTime` is a more informative object, it is also a much more complex one, as it has to deal with issues such as time zones and daylight saving time (for time zone functionalities and conversions, use the external package `TimeZone.jl` (https://github.com/JuliaTime/TimeZones.jl/)).

---

### 2.8.1 Creation of a Date or Time Object ("Input")

There are several ways to obtain the current (local) date/time:

- `todayDate = today()`: Return a `Dates.Date` object.
- `nowTime = now()`: Return a `Dates.DateTime` object.

- `nowTimeUTC = now(Dates.UTC)`: Return a `Dates.DateTime` object with the current UTC time.
- `nowTimeUnix = time()`: Return the so-called "Unix time," a 64-bit integer counting the number of seconds since the beginning of the year 1970.

Alternatively, you can create a `Date` or `DateTime` object from an integer representing the Unix time, or from a string:

- `nowTime = Dates.unix2datetime(nowTimeUnix)`: Where `nowTimeUnix` is as defined in the previous list.
- `christmasLunch = DateTime("2030-12-25T12:30:00", ISODateTimeFormat)`: Return a `DateTime` object from a string in the well-known ISO 8601 format.
- `newYearEvenDinner = DateTime("Sat, 30 Dec 2030 21:30:00", RFC1123Format)`: Return a `DateTime` object from another well-known string format.
- `christmasDay = Date("25 Dec 2030", "d u yyyy")`:
- `newYearDay = Date("2031/01/01", "yyyy/m/d")`:

When parsing the string, you can use the following date and time formatters:

- y: Year digit (e.g., yyyy => 2030, yy => 30)
- m: Month digit (e.g., m => 3, mm => 03)
- u: Name of the month (e.g., "Jan")
- U: Long name of the month (e.g., "January")
- e: Day of the week (e.g., "Tue")
- E: Full name of day of week (e.g., "Tuesday")
- d: Day of the month (e.g., d => 3, dd => 03)

- H: Hour digit (e.g., H => 8, HH => 08)
- M: Minute digit (e.g., M => 0, MM => 00)
- S: Second digit (e.g., S => 0, SS => 00)
- s: Millisecond digit (e.g., .000, fixed 3 digits)

Finally, date and time objects can also be instantiated from a tuple of integers in the following order:

y, m, d, H, M, S, s: - d = Date(2030, 12)

Note that it's not necessary to supply all the values:

dt = DateTime(2030, 12, 31, 9, 30, 0, 0)

## 2.8.2 Extraction of Information from a Date/Time Object ("Output")

You can convert your Date or DateTime object to a string representation with the format function, following the same formatters as listed in the previous section, such as Dates.format(newYearDay, "dd/m/yy") or Dates.format(christmasLunch, "dd/mm/yy H:M:SS").

Otherwise, you can "extract" information using one of the following methods (continuing the example from the previous section):

- year(christmasDay): Return an Int64.
- isleapyear(christmasDay): Return a boolean value (in this case false, as 2030 is not a leap year).
- month(christmasLunch): Return an Int64.
- monthname(christmasDay): Return a String.
- day(christmasDay) (or more verbosely, dayofmonth(christmasDay)): Return an Int64.

- dayname(christmasDay): Return a String.

- dayofweek(christmasDay): Return an Int64 (1 is Monday).

- daysofweekinmonth(christmasDay): Return 4, as there are four Wednesdays in December 2030.

- dayofweekofmonth(christmasDay): Return 4, as December 25, 2030 is the fourth Wednesday of the month.

- hour(christmasLunch): Return an Int64 (also available: minute, second, millisecond).

## 2.8.3 Periods and Date/Time Arithmetic

You can calculate the period between two Date (and the result is a Day object, a subtype of Period) or between two DateTime (and the result is a Millisecond object, another subtype of Period):

- hollidayPeriod = newYearDay - christmasDay

- mealPeriod = DateTime(2030,12,31,23,30) - newYearEvenDinner

You can't mix the two, but you can convert a Date to a DateTime object with convert(DateTime,newYearDay). You could also do the opposite, but you may lose some information.

While the defaults for Date and DateTime arithmetic are days and milliseconds, Julia provides a whole hierarchy of the Period type (you'll see how types can be organized hierarchically in Julia in Chapter 4):

---
- Period
  - DatePeriod
    - Year

- Month
- Week
- Day
- TimePeriod
  - Hour
  - Minute
  - Second
  - Millisecond
  - Microsecond
  - Nanosecond

---

> ❗ **Warning !**
>
> Apart from the risk of an `InexactError`, the ability to convert between these periods is limited. Indeed, periods don't store information about when they start, so how do they determine how many days in a month or in a year? These conversions would be ambiguous. Instead, Julia has another `Period` type, `CompoundPeriod`, which can represent multiple period types.

You can also do arithmetic using a `Date` (or `DateTime`) and a `Period` or between two `Period` objects. In the first case, you would get a new `Date` (or `DateTime`), and in the second case, you would get a `Period` (possibly a `CompoundPeriod`):

- `nextChristmas = christmasDay + Year(1)`: Return a `Date`
- `christmasPresentsOpeningTime = christmasLunch + Hour(3)`: Return a `DateTime`
- `Week(1) + Day(2)`: Return a `CompoundPeriod`

CHAPTER 2   DATA TYPES AND STRUCTURES

You can extract the integer value of a Period with Dates.value(a_period) and extract the individual components of a CompoundPeriod with Dates.periods(a_compounded_period) (the latter requires Julia >= v1.7).

Finally, periods can also be used in ranges:

---

```
semesters = Dates.Date(2020,1,1):Dates.Month(6):Dates.
Date(2022,1,1)
collect(semesters)
```

---

Dates is a large module, and we have covered only the basics here. There is much more (localization, adjustments, and so forth) that you can explore in its official documentation (https://docs.julialang.org/en/v1/stdlib/Dates/).

## 2.9 Memory and Copy Issues

In order to avoid copying large amounts of data, Julia by default copies only the memory address of objects, unless the programmer explicitly requests a so-called "deep" copy or the compiler determines that an actual copy is more efficient.

Use copy() or deepcopy() when you don't want subsequent modifications to the copied object to apply to the original object.

The details are as follows:

- a=b performs a *name binding*, which means it binds (assigns) the entity (object) referenced by b also to the a identifier (the variable name). The possible results are as follows:

    - If b rebinds to some other object, a remains referenced to the original object.

47

CHAPTER 2   DATA TYPES AND STRUCTURES

- If the object referenced by b mutates (i.e., it internally changes), so does (being the same object) those referenced by a.
- If b is immutable and small in memory, under some circumstances the compiler would instead create a new object and bind it to a, but being immutable for the user, this difference would not be noticeable
- As for many high-level languages, you don't need to explicitly worry about memory leaks. A garbage collector exists such that objects that are no longer accessible are automatically destroyed.
- a = copy(b) creates a new, "independent" copy of the object and binds it to a. This new object may in turn reference other objects through their memory addresses, in which case it is their memory address that is copied and not the referenced objects themselves. The possible results are as follow:
  - If the objects referenced by b (e.g., the individual elements of a vector) are rebound to some other objects, the new object referenced by a maintains the reference to the original objects.
  - If the objects referenced by b mutate, so do (being the same objects) those referenced by the new object referenced by a.
- a = deepcopy(b) deep copies everything recursively.

The following code snippet highlights the differences between these three methods of "copying" an object:

CHAPTER 2   DATA TYPES AND STRUCTURES

```
julia> a = [[[1,2],3],4]
2-element Array{Any,1}:
 Any[[1, 2], 3]
 4
julia> b = a
2-element Array{Any,1}:
 Any[[1, 2], 3]
 4
julia> c = copy(a)
2-element Array{Any,1}:
 Any[[1, 2], 3]
 4
julia> d = deepcopy(a)
2-element Array{Any,1}:
 Any[[1, 2], 3]
 4
# rebinds a[2] to an other
  objects.
# At the same time mutates
  object a:
julia> a[2] = 40
40
julia> b
2-element Array{Any,1}:
 Any[[1, 2], 3]
 40
julia> c
2-element Array{Any,1}:
 Any[[1, 2], 3]
 4
julia> c
2-element Array{Any,1}:
 Any[[1, 2], 30]
 4
julia> d
2-element Array{Any,1}:
 Any[[1, 2], 3]
 4
# rebinds a[1][1][2] and at
# the same time mutates a,
# a[1] and a[1][1]:
julia> a[1][1][2] = 20
20
julia> b
2-element Array{Any,1}:
  Any[[1, 20], 30]
 40
julia> c
2-element Array{Any,1}:
 Any[[1, 20], 30]
 4
julia> d
2-element Array{Any,1}:
 Any[[1, 2], 3]
 4
# rebinds a:
julia> a = 5
5
```

*(continued)*

```
julia> d                          julia> b
2-element Array{Any,1}:           2-element Array{Any,1}:
 Any[[1, 2], 3]                    Any[[1, 20], 30]
 4                                 40
# rebinds a[1][2] and at the same julia> c
# time mutates both a and a[1]:   2-element Array{Any,1}:
julia> a[1][2] = 30                Any[[1, 20], 30]
30                                 4
julia> b                          julia> d
2-element Array{Any,1}:           2-element Array{Any,1}:
 Any[[1, 2], 30]                   Any[[1, 2], 3]
 40                                4
```

You can check if two objects have the same values with == and check if two objects are actually the same with === (keep in mind that immutable objects are checked at the bit level and mutable objects are checked for their memory address):

- Given a = [1, 2]; b = [1, 2];, then a == b and a === a are true, but a === b is false.

- Given a = (1, 2); b = (1, 2);, then a == b, a === a, and a === b are all true.

## 2.10 Random Numbers

It is easy to obtain pseudo-random numbers in Julia:

- Random Float64 in [0,1]: rand()

- Random Float32 in [0,1]: rand(Float32)

- Random integer in [a,b]: rand(a:b)

CHAPTER 2   DATA TYPES AND STRUCTURES

- Random float in [a,b] with precision to the second digit: `rand(a:0.01:b)`
- Random float in [a,b]: `rand(Uniform(a,b))` (requires the `Distributions.jl` package, detailed in Chapter 10)
- Random float in [a,b] using a particular distribution (Normal, Poisson, etc.): `rand(DistributionName([distribution parameters]))`

You can obtain an array or a matrix of random numbers by simply specifying the requested size to `rand()`, such as `rand(2,3)` or `rand(Uniform(a,b),2,3)` for a 2×3 matrix.

Sampling uses `Random.GLOBAL_RNG` as the default *random number generator (RNG)* object, but you can have many independent RNGs in your program. An RNG is a "machine" that generates a stream of random numbers. The stream itself is deterministically determined for each "seed" (an integer number) that the RNG has been told to use. Normally this seed changes each time the script is run, so stochastic functions are indeed stochastic and their output will be different each time.

If you want reproducible output, you can fix the seed at the very beginning of your script with

`import Random:seed!; seed!([an_integer])`

Now your script will pick up a specific stream of random numbers, but this stream will always be the same, so the calculations in the script will always return the same result (if the script is run sequentially). If you ever need to revert to nonreproducible randomness, you can use `seed!()` without arguments.

Note that the default Julia RNG only guarantees to provide the same stream of random numbers, conditional on the seed, within minor versions of Julia. If you want to "guarantee" the reproducibility of the output with different versions of Julia, use a package like `StableRNGs.jl`, get an RNG from it, and use that RNG in the stochastic functions instead of the default one:

---
```
using StableRNGs
my_stable_rng = StableRNG(123)
a = rand(my_stable_rng,10)
b = rand(my_stable_rng,10)
my_stable_rng = StableRNG(123)
c = rand(my_stable_rng,10)
b == a # false, it continues the same stream
c == a # true, both are the beginning of identical streams
```
---

## 2.11 Missing, Nothing, and NaN

Julia supports different concepts of *missingness*:

- nothing (type `Nothing`): This is the value returned by code blocks and functions that do not return anything. It is a single instance of the singleton type `Nothing`, and is closer to C-style NULL (sometimes referred to as the "software engineer's null"). Most operations with nothing values will result in a runtime error. In some contexts, it is printed as #NULL.

- missing (type `Missing`): Represents a missing value in a statistical sense—there should be a value, but you don't know which it is (so it is sometimes referred to as the "data scientist's null"). Most operations with missing values will result in missing to propagate (silently). Containers can handle missing values efficiently when they are declared of type `Union{T,Missing}`.

CHAPTER 2  DATA TYPES AND STRUCTURES

- NaN (type Float64): Represents when an operation results in a Not-a-Number value (e.g., 0/0). It is similar to missing in that it propagates silently rather than resulting in a runtime error. Similarly, Julia also offers Inf (e.g., 1/0) and -Inf (e.g., -1/0).

In this book (and in data science in general), we will primarily be dealing with missing values. Julia Base already provides some support for them, such as the functions skipmissing(array) and nonmissing type(Union{T,Missing}). The Missings.jl (https://github.com/JuliaData/Missings.jl) package provides additional methods, such as allowmissing(array) and disallowmissing(array), to convert an Array{T} into an Array{Union{Missing,T}} and vice versa.

## 2.12 Various Notes on Data Types

To convert ("cast") an object into a different type, use convertedObj = convert(T,x). When conversion is not possible, such as when you're trying to convert a 6.4 Float64 into an Int64 value, an error will be thrown (InexactError in this case).

The const keyword, when applied to a variable (e.g., const x = 5), indicates that the identifier cannot be used to bind objects of a different type. The referenced object can still mutate or the identifier rebound to another object of the same type (but in the latter case, a warning is issued). Only global variables can be declared const.

You can "broadcast" a function to work over a collection (instead of a scalar) by using the dot (.) operator. For example, to broadcast parse to work over an array, you would use:

myNewList = parse.(Float64,["1.1","1.2"])

See Chapter 3 for more information about broadcasting.

### 2.12.1 Variable References

Although rarely used, Julia offers the possibility to manually manage the memory of individual variables, in a way similar to the C/C++ pointers. Specifically, y = Ref(x) creates a reference (pointer) to the object bound by x, and y[] dereference representing the pointed object:

```
-----------------------------------------------------------------
x = [1,2,3]
y = Ref(x) # Base.RefValue{Vector{Int64}}([1, 2, 3])
x[2] = 20
y[]        # Vector{Int64}: [1,20,3]
-----------------------------------------------------------------
```

The choice of the square brackets resembles the syntax of the arrays. This is no coincidence, as arrays are always references to their values.

# CHAPTER 3

# Control Flow and Functions

Now that you've been introduced to the main types and containers available in Julia, it is time to put them to work.

## 3.1 Code Block Structure and Variable Scope

Julia supports all the typical flow-control constructs (for, while, if/else, do). The syntax typically takes this form:

```
<keyword> <condition>
    ... block content...
end
```

For example:

```
for i in 1:5
    println(i)
end
```

CHAPTER 3　CONTROL FLOW AND FUNCTIONS

Note that parentheses around the condition are not necessary, and the block ends with the keyword end.

Before I introduce these blocks in detail, let me introduce the concept of scope. The *scope* of a variable is the region of code where the variable can be accessed directly (without using prefixes). Modules, functions, for, and all the other blocks (with the notable exception of if blocks) introduce an inner scope that inherits from the scope where the block is defined (and not, in the case of functions, from the caller's scope).

Variables that are defined outside any block or function are called *global* for the module in which they are defined (the *Main* module if outside of any other module, such as on the REPL), while the others are considered *local*.

Variables defined in a block that already exists as global behave differently depending on whether you are working interactively or not, in particular on assignment.

Take the following example:

```
a = 4
while a > 2
    a -= 1
end
println(a)
```

Here the a variable is global, and a value is assigned to it on each while loop. If you run this snippet in an interactive session, the a variable within the while block is treated as the global variable a, and the snippet works as expected, printing 2. This is to provide convenience when developing interactively, such as on many data exploratory tasks.

However, if you place that snippet in a file and execute or include that file, Julia treats a within the while block as local and prints a warning. In this case, you would also have an error, as the local a is incremented without being first defined. In Julia's terminology, we say that the blocks introduce a *soft scope* that is active only when the code is run noninteractively.

To resolve the ambiguity, you could explicitly declare a as global within the block: global a -= 1 (or, when needed, as local). Even if you don't plan to run your code interactively, this helps to clarify the situation. In general, abusing global variables makes the code more difficult to read, may lead to unintended results, and is often a cause of computational bottlenecks. Use the global variables sparingly!

## 3.2 Repeated Iteration: for and while Loops, List Comprehension, Maps

The for and while constructs are very flexible. First, the condition can be expressed in many ways: for i = 1:2, for i in anArray, while i < 3, and so on. Second, multiple conditions can be specified in the same loop. Consider this example:

```
for i=1:2,j=2:5
 println("i: $i, j: $j")
end
```

In this case, a higher-level loop is started over the range 1:2, and then for each element, a nested loop over 2:5 is executed.

break and continue are supported and work as expected: break immediately aborts the loop sequence, while continue immediately passes to the next iteration.

To loop over the elements of a container, you can

- Loop directly over the elements of the container:

    for element in my_container .... end

- Loop over (a) a flattened version of the indices (eachindex(x)), (b) the incices in a given dimension (axes(x,dim)), or (c) a multidimensional index with CartesianIndex(x):

```
julia> x = [[1,2,3] [4,5,6]]; # 3x2 Matrix
julia> for i in eachindex(x)
           println("$i: $(x[i])")
       end
1: 1
2: 2
3: 3
4: 4
5: 5
6: 6
julia> for c in axes(x,2), r in axes(x,1)
# by col and row
           println("$r,$c: $(x[r,c])")
       end
1,1: 1
2,1: 2
3,1: 3
1,2: 4
2,2: 5
3,2: 6
julia> for i in CartesianIndices(x)
           println("$i: $(x[i])")
       end
CartesianIndex(1, 1): 1
CartesianIndex(2, 1): 2
CartesianIndex(3, 1): 3
```

CHAPTER 3   CONTROL FLOW AND FUNCTIONS

```
CartesianIndex(1, 2): 4
CartesianIndex(2, 2): 5
CartesianIndex(3, 2): 6
```
---

This is preferable to for i in 1:length(x) ... end because x may have custom indices that aren't linear or don't start with 1.

- Loop over the elements and the flattened version of the indices at the same time with

  for (index,element) in enumerate(x) ... end

## ❗ Warning !

Looping over the elements of an array doesn't change them in place:

```
julia> x = [3,2];
julia> function change_me!(x)
         for (i,e) in enumerate(x)
             x[i] = x[i] * 10  # applied
             e = e * 100       # not applied (new binding)
         end
       end
change_me! (generic function with 1 method)
julia> change_me!(x)
julia> x
2-element Vector{Int64}:
 30
 20
```

Julia supports other constructs for repeated iteration, namely list comprehension and maps.

*List comprehension* is essentially a very concise way to write a `for` loop. The following are two examples:

```
[myfunction(i) for i in [1,2,3]]
[x + 2y for x in [10,20,30], y in [1,2,3]]
```

For example, you could use list comprehension to populate a dictionary from one or more arrays:

```
[mydict[i] = value  for (i, value) in enumerate(mylist)]
[students[name] = gender for (name,gender) in zip(names,genders)]
```

You can write complex expressions with list comprehension, such as the following, but at that point perhaps it's best to write the loop explicitly:

```
[println("i: $i - j: $j") for i in 1:5, j in 2:5 if i > j]
```

`map` applies a function to a list of arguments.

The same example of populating a dictionary can also be written (a bit less efficiently) using `map`:

```
map((n,g) -> students[n] = g, names, genders)
```

When mapping a function with a single parameter, the parameter can be omitted, as follows:

```
a = map(f, [1,2,3]) is equal to a = map(x->f(x), [1,2,3])
```

## 3.3 Conditional Statements: if Blocks, Ternary Operator

Conditional statements can be written with the typical `if`/`elseif`/`else` construct:

```
i = 5
if i == 1
    println("i is 1")
elseif i == 2
    println("i is 2")
else
    println("is in neither 1 or 2")
end
```

Multiple conditions can be considered using the logical operators: *and* (&&), *or* (||), and *not* (!) (not to be confused with the bitwise operators & and |).

Note that when the result of the conditional statement can be interfered before the conclusion of the expression evaluation, the remaining parts are not evaluated ("short-circuited"). Typical cases are when the first condition of an && operator is `false` or the first condition of the || operator is `true`: in such cases, it is useless to evaluate the second condition.

If list comprehension is a concise way to write `for` loops, the *ternary operator* is a concise way to write conditional statements. The syntax is as follows:

`a ? b : c`

This means "if a is true, then execute expression b; otherwise, execute expression c." Be sure that there are spaces around the ? and : operators. As with list comprehension, it is important not to abuse the ternary operator in writing too complex conditional logic.

CHAPTER 3   CONTROL FLOW AND FUNCTIONS

## 3.4 Functions

Julia's functions are very flexible. They can be defined inline as f(x,y) = 2x+y, or with their own block introduced using the function keyword:

```
function f(x)
    x+2
end
```

A common third way to define functions is to create an anonymous function and assign this object to a variable.

After a function has been *defined*, you can *call* it to execute it. Note that, as with most high-level languages, there isn't a separate step for *declaring* the function in Julia.

Functions can even be nested, in the sense that the function definition can be embedded within another function definition, or can be recursive, in the sense that there is a call to itself inside the function definition:

```
# A nested function:
function f1(x)
    function f2(x,y)
        x+y
    end
    f2(x,2)
end
# A recursive function:
function fib(n)
    if n == 0   return 0
    elseif n == 1 return 1
    else
```

```
    return fib(n-1) + fib(n-2)
  end
end
```

---

Within the Julia community, it is considered good programming practice for the call to a function to follow these rules:

- To contain all the elements that the function needs for its logic (i.e., no read access to other variables, except constant globals).

- That the function doesn't change any other part of the program that is not within the parameters (i.e., it doesn't produce any "side effects" other than eventually modifying its arguments).

Following these rules will help you achieve code that is fast, reliable (the output of each function depends uniquely on the set of its inputs), and easy to read and debug.

## 3.4.1 Arguments

Function arguments are normally specified by position (*positional arguments*). However, if a semicolon (;) is used in the parameter list of the function definition, the arguments listed after that semicolon must be specified by name (*keyword arguments*).

### ❗ Important !

The function call must respect this distinction, calling positional arguments by position and keyword arguments by name. In other words, it's not possible to call positional arguments by name or keyword arguments by position.

## CHAPTER 3   CONTROL FLOW AND FUNCTIONS

The last arguments (whether positional or keyword) can be specified together with a default value. For example:

- **Definition**: `myfunction(a,b=1;c=2) = (a+b+c)` (definition with two positional arguments and one keyword argument).
- **Function call**: `myfunction(1,c=3)` (calling `(1+1+3)`). Note that b is not provided in the call, and hence the default value is used.

You can optionally restrict the types of the argument the function should accept by annotating the parameter with the type:

`myfunction(a::Int64,b::Int64=1;c::Int64=2) = (a+b+c)`

The reason to specify parameter types is not so much to obtain speed gains. Julia will try to resolve the possible parameter types and return values, and only in rare cases will it be unable to uniquely determine the return type as a function of the input types. In these cases (type-unstable), specifying the parameter types may help solve type instability.

The most important reason to limit the parameter type is to catch bugs early on, when the function is accidentally called with a parameter type it was not designed to work with. In such cases, if the function was annotated with the allowed parameter type, Julia will return to the user a useful error message instead of silently trying to use that parameter.

However, a common case is when you want the function to be called both with single values (scalars) and vectors. You have two options then:

- You can write the function to treat the scalar and rely then on the dotted notation to *broadcast* the function at call time (discussed later in the section "Broadcasting Functions").

- Alternatively, you may want to directly deal with this issue in the function definition. In such cases, you can declare the parameter as being either a scalar type T or a vector T using a Union. For example: function f(par::Union{Float64, Vector{Float64}}) [...] end. You can then implement the logic you want by checking the parameter type using typeof.

Finally, functions in Julia may also accept a variable number of arguments. The splat operator (i.e., the ellipsis ...) can specify a variable number of arguments in the parameter declaration within the function definition and can "splice" a list or an array in the parameters within the function call:

```
myvalues = [1,2,3]
function additionalAverage(init, args...) # The parameter that
                                          uses the ellipsis
                                          must be the
                                          last one
  s = 0
  for arg in args
    s += arg
  end
  return init + s/length(args)
end
a = additionalAverage(10,1,2,3)         # 12.0
a = additionalAverage(10, myvalues ...) # 12.0
```

## 3.4.2 Return Value

Providing a return value using the keyword `return` is optional: by default, functions return the last computed value. Often, `return` is used to immediately terminate a function, for example, upon the occurrence of certain conditions.

Note that the return value can also be a tuple (returning multiple values at once):

```
myfunction(a,b) = a*2,b+2
x,y = myfunction(1,2)
```

## 3.4.3 Multiple-Dispatch

When similar logic should be applied to different kinds of objects (i.e., different types), you can write functions that share the same name but have different types or different numbers of parameters (and different implementations). This highly simplifies the application programming interface (API) of your application, as only one name has to be remembered.

When calling such functions, Julia will pick up the correct one depending on the parameters in the call, selecting by default the stricter version. These different versions are named *methods* in Julia and, if the function is type-safe, dispatch is implemented at compile time and is very fast. You can list all the methods of a given function with `methods(myfunction)`.

The multiple-dispatch polymorphism is a generalization of object-oriented run-time polymorphism in which the same function name performs different tasks depending on the object's class. In traditional object-oriented languages, polymorphism is applied only to a single element, whereas in Julia it applies to all the function arguments (however, object-oriented languages usually have multiple-parameters polymorphism at compile time).

CHAPTER 3   CONTROL FLOW AND FUNCTIONS

We'll go into more depth on multiple-dispatch when dealing with type inheritance in Chapter 4.

## 3.4.4  Templates (Type Parameterization)

Functions can be specified regarding which types they work with. You do this using *templates*:

```
myfunction(x::T, y::T2, z::T2) where {T <: Number, T2} =
x + y + z
```

This function, in the where block, defines two type templates, T and T2, specifying also that T2 must be a child of the type Number, and then the function specifies which of these two types each parameter must be.

You can call the function with (1,2,3) or (1,2.5,3.5) as a parameter, but not with (1,2,3.5), as the definition of myfunction requires that the second and third parameters must be of the same type (whatever that is).

You could write myfunction with the where condition just after the placeholders for the types:

```
myfunction6(x::T where {T <: number}, y::T2 where {T2}, z::T2
where {T2}) = x + y + z
```

In this case, however, you can't use T and T2 as variables in the body of the function, and it takes longer to write if you have multiple template arguments.

## 3.4.5  Functions As Objects

Functions themselves are objects and can be assigned to new variables, returned, or nested. Take this example:

```
---------------------------------------------------------------
f(x) = 2x  # define a function f inline
a = f(2)   # call f and assign the return value to a. `a`
           is a value
```

67

```
a = f      # bind f to a new variable name. `a` is now a
           function
a(5)       # call again the (same) function
```

## 3.4.6 Call by Reference/Call by Value

Julia functions are called using a convention—sometimes known as *call-by-sharing* in other languages—that is somehow in between the traditional *call by reference* (where just a memory pointer to the original variable is passed to the function) and *call by value* (where a copy of the variable is passed, and the function works on this copy).

In Julia, functions work on new local variables, known only inside the function itself. Assigning the variable to another object will not influence the original variable. However, if the object bound with the variable is mutable (e.g., an array), the *mutation* of this object will apply to the original variable as well:

```
function f(x,y)
    x = 10
    y[1] = 10
end
x = 1
y = [1,1]
f(x,y) # x will not change, but y will now be [10,1]
```

(See also the "Memory and Copy Issues" section at the end of Chapter 2.)

CHAPTER 3   CONTROL FLOW AND FUNCTIONS

By convention, functions that change their arguments have ! appended to their name; for example, myfunction!(ref_par, other_pars). The first parameter is, still by convention, the one that will be modified.

## 3.4.7 Anonymous Functions (a.k.a. "Lambda" Functions)

Sometimes you don't need to give a name to a function (e.g., when the function is itself one of the arguments being passed to higher-order functions, like the map function). To define anonymous (nameless) functions, you can use the -> syntax, as follows:

```
x -> x^2 + 2x - 1
```

This defines a nameless function that takes an argument, x, and produces x^2 + 2x - 1. Multiple arguments can be provided using tuples, as follows:

```
(x,y,z) -> x + y + z
```

You can still assign an anonymous function to a variable this way:

```
f = (x,y) -> x+y
```

---

### 🛑 Important !

Do not confuse the single arrow ->, which is used to define anonymous functions, with the double arrow =>, which is used to define a Pair object (e.g., in a dictionary).

---

69

## 3.4.8 Broadcasting Functions

You'll often have a function designed to work with scalars that you want to apply repetitively to values within a container, like the elements of an array. Instead of writing for loops, you can rely on a native functionality of Julia, which is to *broadcast* the function over the elements you wish. Take, for example, the following function:

f1(a::Int64,b::Int64) = a*b

It expects as input two scalars, such as f1(2,3). But what if a and b are vectors (say a=[2,3] and b=[3,4])? You can't directly call the function f1([2,3],[3,4]).

The solution is to use the function broadcast(), which takes the original function as its first argument followed by the original function's arguments: broadcast(f1,[2,3],[3,4]). The output is a vector that holds the result of the original function applied first to (a=2,b=3) and then to (a=3,b=4).

A handy shortcut to broadcast() is to use *dot notation*, which is the original function name followed by a dot: f1.([2,3],[3,4]).

Sometimes the original function takes natively some parameters as vector, and you want to limit the broadcast to the scalar parameters. In such cases, you can use the Ref() function to protect the parameters that you don't want to be broadcast:

---
```
f2(a::Int64,b::Int64,c::Array{Int64,1},d::Array{Int64,1})
 = a*b+sum(c)-sum(d)
f2(1,2,[1,2,3],[0,0,1]) # normal call without broadcast
f2.([1,1,1],[2,2,2],Ref([1,2,3]),Ref([0,0,1])) # broadcast over
the first two arguments only
```
---

CHAPTER 3   CONTROL FLOW AND FUNCTIONS

## 3.5  do Blocks

In Julia, do blocks allow developers to define "anonymous" functions that are passed as the first argument of the outer functions. For example, you write

f1(f2,x,y) = f2(x+1,x+2)+y

To use this f1 function, you first need another function to act as f2 (the inner function). You could define it as f2(g,z) = g*z and then call f1 as f1(f2,2,8). Every time the first argument is a function, this can be written with a do block. Hence, you can obtain the same result using the do block:

```
f1(2,8) do i,j
    i*j
end
```

With the do block, at the same time you call f1 and define an anonymous function to act as f2. i and j are local variables that are made available to the do block. Their values are determined in the f1 function (in this case, i=2+1 and j=2+2). The result of the block computation is then made available as the output of the function acting as the first parameter of f1. Again, what you do with this value is specified in the definition of the f1 function (in this case, the value 8 is added to it to form 20, the returned value).

Another typical use of do blocks is within input/output operations, as discussed in Chapter 5.

## 3.6 Exiting Julia

To exit a running Julia session, press Ctrl+D (or use Ctrl+C to throw a InterruptException when possible, regaining control of the Julia prompt without closing the process).

To exit the script programmatically, use exit(exit_code), where the default exit code of zeros indicates, by convention, a normal program termination.

Finally, sometimes you'll want to be able to define something your code should run whenever the Julia process exits. This is exactly the role of the atexit(f) function, which allows you to define a zero-argument function f() to be called upon exit. You can use it, for example, to start a new "clean Julia session" when a given function is invoked. Type the following function in your global or local startup.jl file (e.g., ~/.julia/config/startup.jl in Linux):

```
function workspace()
    atexit() do
        run(`$(Base.julia_cmd())`)
    end
    exit()
end
```

Now, whenever the workspace() function is invoked, the current Julia process will be closed and a new one will be started. (Note that workspace was the name of a function in previous Julia versions that cleaned up the workspace, but that is no longer available in recent versions. This script replicates its functionality, albeit at the cost of forcing a restart of the whole Julia process.)

# CHAPTER 4

# Custom Types

In Chapter 2, we discussed built-in types, including containers. In this chapter, we explore how to create user-defined types.

---

 **Note**

**"Type" vs. "Structure"**

Let's clarify these two terms in the Julia language context. A *type* of an object, in plain English, refers to the set of characteristics that describe the object. For example, an object of type sheet can be described with its dimensions, its weight, its size, or its color. All values in Julia are true objects belonging to a given type (they are individual *instances* of the given type). Julia types include *primitive types*, which include objects made of a fixed number of bits (e.g., all numerical types such as Int64, Float64, and Char), and *composite types*, also known as *structures*, where the object's characteristics are described using multiple fields and a variable number of bits. Both structures and primitive types can be user-defined and are hierarchically organized. Structures roughly correspond to what are known as *classes* in other languages.

---

© Antonello Lobianco 2024
A. Lobianco, *Julia Quick Syntax Reference*, https://doi.org/10.1007/979-8-8688-0965-1_4

Two operators are particularly important when working with types:

- The `::` operator is used to constrain an object to being of a given type. For example, `a::B` means "a must be of type B."

- The `<:` operator is used to indicate that the type on the left of the operator is (or has to be) a subtype of the type on the right. For example, `A<:B` means "A must be a subtype of B"; that is, B is the "parent" type and A is its "child" type.

The primary reason to use these operators is to confirm that the program works the way it is expected or to create code that specializes on particular types. In some cases, providing the type information also improves performance.

## 4.1 Primitive Type Definition

A user-defined primitive type is defined with the keyword `primitive type`, followed by its name and the number of bits it requires:

```
primitive type [name] [bits] end
```

For example:

```
primitive type My10KBBuffer 81920 end
```

**Caution**

A current limitation of primitive types is that the number of bits must be a multiple of 8 and below 8388608.

Optionally, a parent type can be specified as follow:

```
primitive type [name] <: [supertype] [bits] end
```

Note that the internal representation of two user-defined types with the same number of bits is exactly the same. The only thing that would change is their names, but that's an important difference: it's the way that functions act when objects of these types are passed as arguments that changes; in other words, it's the use of named types across the program that distinguishes them, rather than their implementation.

## 4.2 Structure Definition

To define a structure instead of a primitive type, you use the keyword `mutable struct`, give the structure a name, specify the fields, and close the definition with the end keyword:

```
mutable struct MyOwnType
  field1
  field2::String
  field3::Int64
end
```

Note that while you can optionally define each individual field to be of a given type (e.g., `field3::Int64`), you can't define fields to be subtypes of a given type (e.g., `field3 <: Number`). In order to do that, you can use templates in the structure definition:

```
mutable struct MyOwnType2{T<:Number}
 field1
 field2::String
 field3::T
end
```

Using templates, the definition of the structure is dynamically created the first time an object whose field3 is of type T is constructed.

The type with which you annotate the individual fields can be either a primitive one (like in the preceding example) or a reference to another structure (an example is provided later in the chapter).

Another difference from other high-level languages (e.g., Python) is that in Julia you can't add or remove fields from a structure after you have defined it. If you need this functionality, you must use dictionaries instead, but be aware that you will trade this flexibility for worse performance.

Conversely, to gain performance (but sacrifice flexibility), you can omit the mutable keyword in front of struct. That way, once an object of that type has been created, its fields can no longer be changed (i.e., structures are immutable by default). Note that mutable objects (as arrays) remain themselves mutable even in an immutable structure.

To define defaults for some fields, either use a constructor (introduced in the next section) or prepend the @kwdef keyword to the structure definition:

```
@kwdef mutable struct MyOwnType3
  field1            = "foo"
  field2::String
  field3::Int64     = 1
end
```

## 4.3 Object Initialization and Usage

After you define your custom type (primitive or structure), you can initialize as many objects as you wish from that type:

```
my_object = MyOwnType("something","something",10)
a = my_object.field3  # 10
my_object.field3 = 20 # only if my_object is a mutable struct
```

Note that you have to instantiate the object with the values in the same order specified in the structure definition. You can't instantiate objects by name (rather, consider named tuples for that):

```
MyOwnType(field3 = 10, field1 = "something", field2 = "something") # Error!
```

For types defined with @kwarg, in addition to the usual position-based object instantiation, there is a keyword-based method where all fields must be provided with keywords, and only the nondefault ones are mandatory:

```
my_object = MyOwnType3(field2="bbb",field3=10)
```

If you need to apply logic during object initialization—for example, to validate input, set defaults, or calculate one field as a function of the others—you would use *constructors*, which are simply functions of the same name as the type. They can be defined inside the structure definition (inner constructor) or outside it. The object instantiation you just saw is actually a call to the default position-based constructor that is automatically created when you define a struct.

For example, you may want to consider an external constructor for MyOwnType as follows:

```
function MyOwnType(i::T) where {T <: Number}
  i >= 0 || @error "Only non-negative numbers, please
  (you provided $i)"
  return MyOwnType(sqrt(i),string(i),i)
end
```

Similarly, you could consider an internal constructor for `MyOwnType2`, to be defined inside the structure definition of `MyOwnType2`, as follows:

```
mutable struct MyOwnType2{T<:Number}
  field1
  field2::String
  field3::T
  function MyOwnType2(i::T) where {T <: Number}
    i >= 0 || @error "Only non-negative numbers, please (you
    provided $i)"
    return new{T}(sqrt(i),string(i),i)
  end
end
```

Inner constructors *replace* the default constructor, while external constructors are just additional functions. So you can now instantiate objects as shown here:

```
MyOwnType(3,"bbb",3) # default position-based constructor
MyOwnType(3)         # alternative external constructor
MyOwnType2(3)        # custom inner constructor
```

But this would produce an error, because for `MyOwnType2` you no longer have the default positional constructor:

`MyOwnType2(5,"aaa",3) # error`

To access object fields or change their value in Julia, you use, like in most other languages, the `object.field` syntax.

**Tip**
**Functions and Structures**

Note that functions that deal with a certain object are not defined or declared within that type definition; that is, they are not associated with one specific type. This implies that instead of calling an object method, such as `myobj.func(x,y)`, in Julia you would pass the object as a parameter, such as `func(myobj,x,y)`, no matter whether it's as a first, second, or subsequent argument.

## 4.4 Abstract Types and Inheritance

You can create abstract types using the keyword `abstract type`. Abstract types are not allowed to hold any field, and objects cannot be instantiated from them. Rather, concrete types with fields can be defined as subtypes of them and objects can be instantiated from them. In turn, concrete types can't have subtypes.

You can create a whole hierarchy of abstract types, although multiple inheritance (when a type is a subtype of multiple types) is not supported:

```
abstract type MyOwnGenericAbstractType end
abstract type MyOwnAbstractType1 <: MyOwnGenericAbstractType end
abstract type MyOwnAbstractType2 <: MyOwnGenericAbstractType end
mutable struct AConcreteType1 <: MyOwnAbstractType1
    f1::Int64
    f2::Int64
end
```

```
mutable struct AConcreteType2 <: MyOwnAbstractType1
  f1::Float64
end

mutable struct AConcreteType3 <: MyOwnAbstractType2
  f1::String
end
```

Why would you need inheritance if abstract types cannot have fields? It's not to save the time of defining common fields across multiple types at once (you can use *composition* for that, discussed in the next section). Rather, it is again in the *usage* of types across the program that the definition of their hierarchical structure becomes useful.

Consider the following objects:

```
o1 = AConcreteType1(2,10)
o2 = AConcreteType2(1.5)
o3 = AConcreteType3("aa")
```

By the fact they are all subtypes of `MyOwnGenericAbstractType`, you can define a function that provides a *default* implementation for them:

```
function foo(a :: MyOwnGenericAbstractType)
  println("Default implementation: $(a.f1)")
end
foo(o1) # Default implementation: 2
foo(o2) # Default implementation: 1.5
foo(o3) # Default implementation: aa
```

You can then decide to provide a function to offer a more specialized implementation for all objects whose type is a subtype of MyOwnAbstractType:

```
function foo(a :: MyOwnAbstractType1)
    println("A more specialised implementation: $(a.f1*4)")
end
foo(o1) # A more specialised implementation: 8
foo(o2) # A more specialised implementation: 6.0
foo(o3) # Default implementation: aa
```

This is possible thanks to the *multiple-dispatch* mechanism discussed in Chapter 3: when multiple *methods* of a function are available to dispatch a function call, Julia will choose the stricter one—the one defined over the exact parameter's types or their more immediate *supertypes* (parent types):

```
function foo(a :: AConcreteType1)
    println("A even more specialised implementation: $(a.f1 + a.f2)")
end
foo(o1) # A even more specialised implementation: 12
foo(o2) # A more specialised implementation: 6.0
foo(o3) # Default implementation: aa
```

While the preceding example has a single argument, "multiple" refers to the dispatch mechanism that can be applied, in a similar way, to any number of functional arguments.

> **❗ Important !**
>
> One point of attention is when you want to use inheritance for parametric types. While both `Vector{Int64} <: AbstractVector{Int64}` and `Int64 <: Number` are true, it is false that `AbstractVector{Int64} <: AbstractVector{Number}` or that `Vector{Int64} <: Vector{Number}`.
>
> If you want to allow a function parameter to be a vector of numbers, use instead templates explicitly; for example:
>
> `foo(x::AbstractVector{T}) where {T<:Number} = return sum(x)`

### 4.4.1 Implementation of the Object-Oriented Paradigm in Julia

Julia allows both inheritance and composition models, although with different levels of support.

*Inheritance*, as you just saw, is when a hierarchical structure of types is obtained by declaring one type as subtype of another. You also saw the limits of using inheritance in Julia, when only the behavior (and not the structure) can be inherited. So how do you implement an object-oriented model in Julia? The preference in the Julia community is to use composition over inheritance.

*Composition* is when you declare one field of a given type as being an object of another composite type. Through this (referenced) object, you then gain access to the fields of the other type.

Consider the following example, which first defines a generic `Person` structure and then defines two more specific `Student` and `Employee` structures:

---

```
struct Shoes
    shoesType::String
    colour::String
end

struct Person
    myname::String
    age::Int64
end

struct Student
    p::Person
    school::String
    shoes::Shoes
end

struct Employee
    p::Person
    monthlyIncomes::Float64
    company::String
    shoes::Shoes
end
```
---

Instead of using inheritance to declare `Student` and `Employee` as subtypes of `Person`, this example uses composition to assign a field p of type `Person` to both of them. It is thanks to this field that you do not need to repeat the fields that are common to both.

## CHAPTER 4  CUSTOM TYPES

---

 **Important !**

Types must be defined before they can be used to reference objects. Hence, the Shoes definition must come before the Students and Employee definitions.

---

You can then create instances of the specialized type, either by creating the referenced object first or by doing that inline in the constructor of the specialized type:

```
gymShoes = Shoes("gym","white")
proShoes = Shoes("classical","brown")

Marc = Student(Person("Marc",15),"Divine School",gymShoes)
MrBrown = Employee(Person("Brown",45),3200.0,"ABC Corporation
Inc.", proShoes)
```

Finally, you can use multiple dispatch to provide tailored implementation for the specialized types and access referenced objects and/or general fields through a chained use of the dot (.) operator:

```
function printMyActivity(self::Student)
   println("Hi! I am $(self.p.myname), I study at
   $(self.school) school, and I wear $(self.shoes.colour) shoes")
end

function printMyActivity(self::Employee)
  println("Good day. My name is $(self.p.myname), I work at
  $(self.company) company and I wear $(self.shoes.colour) shoes")
```

**end**

```
printMyActivity(Marc)      # Hi! I am Marc, ...
printMyActivity(MrBrown)   # Good day. My name is MrBrown, ...
```

---

While using inheritance wisely can in practice suffice for many modeling design situations you may encounter, it is still true that the Julia core language misses some expressiveness in the sense that you cannot directly consider and/or distinguish between different concepts of relations between objects, like *specialization* (e.g., Person→Student), *composition* (e.g., Person→Arm), and *weak relation* (e.g., Person→Shoes).

To address this lack of expressiveness, several third-party packages (not discussed in this book) have been released that improve Julia flexibility in this area, like the `SimpleTraits.jl` (https://github.com/mauro3/SimpleTraits.jl) package to imitate multiple inheritance and the `OOPMacro.jl` (https://github.com/ipod825/OOPMacro.jl) package to automatically copy field declaration from a parent to a child type.

## 4.5 Some Useful Functions Related to Types

To complete the discussion concerning user defined types, you may find the following type-related functions useful:

- `supertype(MyType)`: Return the parent types of a type.
- `subtypes(MyType)`: List all children of a type.
- `fieldnames(MyType)`: Query all the fields of a structure.
- `isa(obj,MyType)`: Check if `obj` is of type `MyType`.
- `typeof(obj)`: Return the type of `obj`.
- `eltype(obj)`: Return the inner type of a collection, like the inner elements of an array.

## 4.6 Definition of Common Julia Terms

The Julia jargon about types sometimes can be intimidating. To help you navigate it comfortably, Table 4-1 summarizes the main definitions:

*Table 4-1. Common Julia Terms*

| Term | Definition |
| --- | --- |
| Primitive type | A type defined with the `primitive type` keyword. Objects of a primitive type have a given fixed memory size specified in the type definition. |
| Composite type | A type defined with the `struct` keyword. Composite types are formed by zero or more fields referencing other objects (of primitive or composite type). |
| Singleton | An object instantiated from a composite type formed by zero fields. |
| Abstract type | A type defined with the `abstract type` keyword. Abstract types have no fields and objects cannot be instantiated from them. Furthermore, they cannot be declared a child of a concrete type. |
| Concrete type | A primitive or composite type. |
| Mutable type | A composite type defined with the `mutable struct` keyword. Mutable types can have their fields rebound to other objects than those associated at the time of initialization. |
| Immutable type | All types except those defined with `mutable struct`. |
| Parametric type | A family of (mutable or immutable) composite or abstract types with share the same field names and type names, but not the parameters' types. The specific individual type is then uniquely identified with the name of the parametric type and the type(s) of the parameter(s). |

*(continued)*

*Table 4-1.* (*continued*)

| Term | Definition |
|---|---|
| Container or collection | A composite type (not necessarily mutable) that is designed to reference a variable number of objects and provides methods to access, iterate, and eventually mutate the reference to other objects. |
| Predefined type | A type whose definition is provided in Julia Base or in the Julia Standard Library. |
| Bits type | A primitive or immutable composite type whose fields are all bits type themselves. |
| Trait | A type used to group other types outside the default type hierarchy and create functions that dispatch over this trait. |

# 4.7 Exercise 1: The Schelling Segregation Model

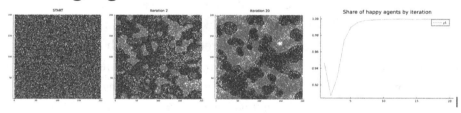

In this exercise you will implement Schelling's model of segregation, (https://en.wikipedia.org/wiki/Schelling%27s_model_of_segregation) a classic agent-based model in the social sciences, which introduces the concepts of emergent macro-behaviors and tipping points, and which led to its author (Thomas Schelling) being awarded the Nobel Prize in Economics in 2005. Schelling's model was one of the earliest models to present results obtained from running simulations, especially in the social sciences.

## CHAPTER 4 CUSTOM TYPES

The world is modeled as a gridded space inhabited by two groups (in the early 1970s, racial and segregation issues concerning the co-existence of "blacks" and "whites" were topical). Each member of the group has a preference to live in a neighborhood inhabited by agents of its own type, with a certain "tolerance." If the proportion of agents of its own type in the neighborhood falls below this "tolerance" level, the agent will try to move to a place with a higher proportion of agents of its own type.

The main finding of the paper published by Schelling[1] was that even a relatively mild preference for the presence of similar agents would have led to a segregated world, and that there was a specific threshold (between 30% and 40%) that drove these two completely different outcomes—that is, a "tipping point."

The simulation algorithm works like this: at each step, it looks at each agent, checks whether it is "happy" with its current location (looking at the proportion of its own types in the neighborhood), and, if not happy, moves the agent to a position where it would be happy (setting its previous location as empty).

There are several "generality" vs. "specificity" ways to code the algorithm. One way would be to hard-code the two agent types, such as `blue` and `red`. Another way would be to take a very generic approach and create an abstract type `agent` and a concrete class for each agent type. The skeleton exercise presented in this section takes an intermediate approach, with only one `Agent` class and the type of agent encoded as an integer, where 0 represents an empty cell. Feel free to use it or develop your own algorithm from scratch.

**Skills used:** designing custom types; defining and calling functions; manipulating arrays; using conditional statements and `for` loops; plotting.

---

[1] Thomas C. Schelling (1971), "Dynamic models of segregation," *The Journal of Mathematical Sociology*, 1:2, 143–186, DOI: https://doi.org/10.108 0/0022250X.1971.9989794.

## 4.7.1 Instructions

The skeleton of the exercise presented next includes some code that is already set up, which you must complete by replacing [...] Write your code here with your own code in Step 4. If you want to avoid typing, you can find this skeleton in the GitHub repository of the book, (https://github.com/Apress/Julia-Quick-Syntax-Reference-2nd-ed) where you will also find its solution, including a version that uses Agents.jl, (https://github.com/JuliaDynamics/Agents.jl) a powerful framework to implement agent-based models in Julia.

## 4.7.2 Skeleton

### 4.7.2.1 STEP 1: Set up the environment…

```
cd(@__DIR__)
using Pkg
Pkg.activate(".")
# If you are using a Julia version different than 1.11 please
uncomment and run the following 2 lines (reproducibility
guarantee will however be lost)
# Pkg.up()
# Pkg.resolve()
Pkg.instantiate()
using Random, Plots
Random.seed!(123)
```

## 4.7.2.2 STEP 2: Define the Agent and Env classes...

```
mutable struct Agent
    gid::Int64
end
mutable struct Env
    nR::Int64                          # number of rows
    nC::Int64                          # number of columns
    similarityThreshold::Float64       # threshold for agents to
                                       # be "happy" with their
                                       # location
    neighborhood::Int64                # how far looking for
                                       # "similar" agents
    nSteps::Int64                      # number of interactive
                                       # steps to employ
    cells::Vector{Agent}               # total cells in the
                                       # environment
    gids::Vector{Int64}                # ids of the agents types
                                       # (or "groups")
    grsizes::Vector{Int64}             # number of agents
                                       # per group
end
```

## 4.7.2.3 STEP 3: Define some utility functions...

```
xyToId(x,y,nR,nC)   =   nR*(x-1)+y
iDToXY(id,nR,nC)    =   Int(floor((id-1)/nR)+1), (id-1)%(nR)+1
printableGrid(env) = reshape([a.gid for a in env.cells],env.
nR,env.nC)
```

CHAPTER 4   CUSTOM TYPES

## 4.7.2.4  STEP 4: Define the main functions of the algorithm...

```
"""
    getNeighbours(x,y,env,gid=nothing)
```
Return the number of total neighbours if `gid` is `nothing` (skipping the empty cells), or a specific `gid` if one is given.
```
"""
function getNeighbours(x,y,env,gid=nothing)
    board  = reshape(env.cells,env.nR,env.nC)
    region = board[max(1,y-env.neighborhood):min(nR,y+env.neighborhood),max(1,x-env.neighborhood):min(nC,x+env.neighborhood)]
    # [...] Write your code here
    # You need to find and sum the elements in the `region`
      array that match the gid or that are not empty
end
"""
    isHappy(x,y,a,env)
```
Return whether the specific agent `a` is happy at his current location
```
"""
function isHappy(x,y,a,env)
    # [...] Write your code here
    # You need to compare the number of total neighbours (all
      gids) and compare with the number of neighbours of the
      specific gid of the `a` agent (`a.gid`). Finally check
      this with the similarity threshold specified in the `env`
      structure and return a boolean value.
```

## CHAPTER 4 CUSTOM TYPES

**end**
"""
    reallocatePoints!(env)

Loop over all the cells and if an agent on that location is unhappy, it moves it to a location where it is happy and set the departing cell as empty (i.e. occupied by an agent whose gid is zero).
It returns the share of agents that were happy before the move.
"""
**function** reallocatePoints!(env)
    happyCount = 0
    **for** (i,a) **in** enumerate(env.cells)
        # [...] Write your code here
        # You need to check if there is an unhappy agent in this specific cell, and if that is the case, loop again to find a suitable cell and move the agent there.
        # "moving" is implemented by changing the destination cell gid to the gid of the incoming agent, and setting its original cell as empty
    **end**
    **return** happyCount/sum(env.grsizes)
**end**
"""
    modelrun!(env)

Run the reallocation algorithm for the given steps printing a heatmap at each iteration.
Also, print at the end the chart of the happy agents by epoch
"""
**function** modelrun!(env)

```
    outplot = heatmap(printableGrid(env), legend=nothing,
    title="START", color=mypal,aspect_ratio=env.nR/env.nC,
    size=(600,600*env.nR/env.nC))
    nHappyCount = Float64[]
    display(outplot)
    for i in 1:env.nSteps
        println("Running iteration $i...")
        nHappy = reallocatePoints!(env)
        push!(nHappyCount,nHappy)
        outplot = heatmap(printableGrid(env), legend=nothing,
        title="Iteration $i", color=mypal,aspect_ratio=env.nR/
        env.nC, size=(600,600*env.nR/env.nC))
        display(outplot)
    end
    happyCountPlot = plot(nHappyCount,title="Share of happy
    agents by iteration")
    display(happyCountPlot)
end
```

---

### 4.7.2.5 STEP 5: Set the parameters of the specific simulation to run…

---

```
# Parameters...
nR        = 200
nC        = 200
nSteps    = 20
similarityThreshold = 0.4          # Agent is happy if at least
                                   40% similar
neighborhood = 5                   # Defining how far looking
                                   for similar agents
```

```
mypal         = [:white,:red,:blue]   # First color is for the
                                        empty cell
gids          = [1,2]                 # Gid 0 is reserved for
                                        empty cell
grShares      = [0.4,0.4]             # Shares of cells occupied
                                        by agents, by type
```

## 4.7.2.6 STEP 6: Initialize the simulation with the given parameters...

```
nCells  = nR*nC
nGroups = length(gids)
grsizes = Int.(ceil.(nCells .* grShares))

cells   = fill(Agent(0),nCells)
count = 1
for g in 1:nGroups
    [cells[j] = Agent(gids[g]) for j in
count:count+grsizes[g]-1]
    count += grsizes[g]
end

shuffle!(cells)
env = Env(nR,nC,similarityThreshold,neighborhood,nSteps,cells,
gids,grsizes)
heatmap(printableGrid(env), legend=nothing, title="Iteration
0", color=mypal,aspect_ratio=env.nR/env.nC, size=(600,600*env.
nR/env.nC))
```

### 4.7.2.7 STEP 7: Run the model...

```
modelrun!(env)
```

## 4.7.3 Results

If you've followed the instructions correctly, you should have a series of graphs where the red and blue pixels are random at first, but then start to "aggregate" and form clusters of regions of a single color, until an equilibrium is found where all agents are happy and no longer move.

## 4.7.4 Possible Variations

You can try variations on this model. Some ideas include

- Implement several groups of agents (a version of this is available in the book's repository).

- Implement a version that uses a framework designed to facilitate agent-based modeling (a version that uses the `Agents.jl` package is available in the book's repository).

- Run Monte Carlo simulations of the tolerance threshold, where it is a property of the agent rather than the group, perhaps with the group having some distribution of it and each agent in that group sampling from it.

- Implement two thresholds, one to define happiness and one to define the actual decision to move (to take into account the costs associated with relocation).

# CHAPTER 5

# Input/Output

The following third-party packages are covered in this chapter:

| CSV.jl | https://github.com/JuliaData/CSV.jl | v0.10.14 |
| HTTP.jl | https://github.com/JuliaWeb/HTTP.jl | v1.10.8 |
| XLSX.jl | https://github.com/felipenoris/XLSX.jl | v0.10.1 |
| OdsIO.jl | https://github.com/sylvaticus/OdsIO.jl | v0.6.3 |
| JSON3.jl | https://github.com/quinnj/JSON3.jl | v1.14.0 |

Input/output (IO) in Julia is implemented by first choosing the appropriate `IOStream` object—for example, a file, the user terminal, or a network object—and then applying the desired function over that stream. The general idea is to open the stream, perform the required operations, and then close the stream.

Concerning the user's terminal, Julia provides interaction with it through the built-in `stdin` and `stdout` (and eventually `stderr`) IO streams. These are already opened and are the default streams for input and output operations.

Concerning web resources, you can use the Standard Library for basic downloads, but for more advanced operations you need to use a third-party package. We will consider `HTTP.jl` (https://github.com/JuliaWeb/HTTP.jl) in this chapter.

CHAPTER 5   INPUT/OUTPUT

# 5.1  File System Functions

Before we look at the various methods for handling IO in Julia, let me introduce you to the functions Julia provides for interacting with the file system.

Julia allows you to run OS-specific commands by simply using the backticks to define a command and run to execute it. For example, in Linux you could run

---
```
dir_to_list    = "/etc"
command_to_run = `ls $dir_to_list`
run(command_to_run)
```
---

However, using these commands in a script would make the script OS-specific. Instead, Julia provides functions that are inspired by the Unix systems but work in an OS-independent way. You can browse the full set of functions that deal with the file system in the relevant section (https://docs.julialang.org/en/v1/base/file/) of the documentation. Here I list the most common ones:

- pwd(): Print the current directory.
- readdir("a/dir"): Return a vector of items in the given directory. See also walkdir to recursively walk through a given directory.
- ispath("foo"): Return whether the given argument (as a string) is an existing path. There are also the more specific isfile("test.jl") and isdir("test").
- isabspath("/path/to/file.txt"): Return whether the given argument (whether present or not) is an absolute path.

CHAPTER 5  INPUT/OUTPUT

- `basename("/path/to/file.txt")`: Return the filename part of the given path (i.e., "file.txt").
- `dirname("/path/to/file.txt")`: Complement `basename` to return the directory name of the given path (i.e., "/path/to")
- `abspath("path/to/foo.txt")`: Return the absolute path of the specified path (whether it exists or not). Equivalent to `joinpath(pwd(),"path/to/foo.txt")`.
- `realpath("../path/to/some/file.txt")`: Return the absolute path to an existing file or directory.
- `cd(@__DIR__)`: Set the working directory to the given one; in this example, the directory of the current file.
- `mkdir("foodir")`: Create a directory.
- `mkpath("goodir/foo/baa")`: Create a directory with all necessary intermediate directories.
- `joinpath("aaa","bbb")`: Join the different subpaths, taking into account the different OS formats. For example, to write OS-independent code, use `my_file=joinpath("repository","data.txt")` instead of `my_file="repository/data.txt"`.
- `rm("foo.txt",force=true,recursive=true)`: Remove a file system item. Don't raise an error if it doesn't exist, and do the removal recursively if the item is a non-empty directory.
- `mv("oldpath","newpath")`: Move (rename) a file system item from the old path (source) to a new path (destination).
- `cp("oldpath","alternativepath")`: Copy a file system item from the old path to an alternative path.

99

CHAPTER 5   INPUT/OUTPUT

## 5.2  Reading (Input)
### 5.2.1  Reading from a Terminal

To read user input from a terminal, you use `readline`:

- `aString = readline()`: Read whatever the user types into the terminal until the user presses Enter.
- `anInteger = parse(Int64, readline())`: Read a number entered by the user in the terminal.

Note that you will rarely want to read user input from a terminal. Julia is interactive, so you can run portions of code or change the variables directly from the terminal or from the editor that you are using. Still, you can create a Julia script (like in the following example) and run it with either `julia myScript.jl` or with `include("myScript.jl")` from within the Julia prompt. This will mimic the classic "scripts" behavior, where the user runs a script without having full control over it, but only entering the user input that is requested by the script:

---
```julia
println("Welcome to a julia script.")
if length(ARGS)>0
  println("You used this script with the following arguments:")
  for arg in ARGS
    println("- $arg")
  end
end
function getUserInput(T=String,msg="")
  print("$msg ")
  if T == String
      return readline()
```

```
    else
      try
        return parse(T,readline())
      catch
        println("Sorry, I could not interpret your answer. Please
        try again")
        getUserInput(T,msg)
      end
    end
end
sentence = getUserInput(String,"Which sentence do you want to
be repeated?");
n        = getUserInput(Int64,"How many times do you want it to
be repeated?");
[println(sentence) for i in 1:n]
println("Done!")
```

Note that in this code snippet, the `getUserInput` function uses runtime exceptions to be sure that the user's response is compatible with the type you want. You'll learn more about runtime exceptions in Chapter 8. Command-line arguments are available through the variable `ARGS`. Use the `ArgParse` (https://github.com/carlobaldassi/ArgParse.jl) package for parsing complex command-line options.

## 5.2.2 Reading from a File

File reading (and writing, as you will see) is similar in Julia to other languages. You open the file, specify the modality (`r`, which stands for "reading"), bind the file to an object, operate on the object, and close the file when you are done. A better alternative is to encapsulate the file operations in a do block that closes the file automatically when the block ends.

### 5.2.2.1 Read the Whole File in a Single Operation

---
```
open("afile.txt", "r") do f   # "r" for reading
  filecontent = read(f,String) # attention that it can be used
  only once. If used a second time, without reopening the file,
  read() would return an empty string
  # ...operate on the whole file content all at once...
end
```
---

### 5.2.2.2 Read the File Line by Line

---
```
open("afile.txt", "r") do f
  for ln in eachline(f)
    # ... operate on each individual line at a time...
  end
end
```
---

This method can be used even with very large files that don't fit into memory.

### 5.2.2.3 Parsing Comma-Separated Value (CSV) Files

For reading comma-separated value (CSV) files—and in general, any textual delimited files—you can simply use the built-in readdlm() function available from the Standard Library package DelimitedFiles.jl.

You can skip rows and/or columns using the slice operator and then convert to the desired type:

```
my_data = convert(Array{Int64},readdlm("myinputfile.csv",','))
[2:end,3:end]) # Skip the first 1 row and the first 2 columns
```

CHAPTER 5  INPUT/OUTPUT

readdlm output is a standard Array. If you want more options or need the data in a different format, you can use the read function from the CSV.jl (https://github.com/JuliaData/CSV.jl) package:

my_data = CSV.read(file, destination_sink; [options])

where destination_sink is a function that supports the Tables.table https://github.com/JuliaData/Tables.jl interface. Use CSV.Tables.matrix as the destination sink to obtain the data as a standard Matrix, or DataFrame for a widely used tabular format (which we will explore extensively in Chapter 9).

CSV.read supports a long list of (mostly self-explanatory) options: header (unless this option is set to false, the first row would be interpreted as the headers), normalizenames, skipto, footerskip, transpose, comment, ignoreemptyrows, select/drop, limit, buffer_in_memory, ntasks, rows_to_check, missingstring, delim (use '\t' for tab-delimited files), ignorerepeated, quoted, quotechar/openquotechar/closequotechar, escapechar, dateformat, decimal/groupmark, truestrings/falsestrings, stripwhitespace, types, typemap, pool, downcast, stringtype, strict/silencewarnings/maxwarnings, debug, validate.

The type of each field is auto-recognized using a large number of initial rows, but sometimes this is not enough. For example, in some datasets you may have a field with all missing values and then suddenly non-missing values appear very late in the dataset. Such situations would trigger a TypeError when you would try to populate the field with some values, as it would have been recognized of type Missing. The trick is to manually specify the column type with the types parameter (a vector or a dictionary, such as types=Dict("myFieldFoo" => Union{Missing,Int64})). Note that CSV.read can be used with inputs other than files. You'll see an example later in this chapter.

## 5.2.3 Importing Data from Excel

Importing data from Excel can be accomplished by using the XLSX.jl package, which enables you to do the following:

- **Retrieve a list of the sheet's names**: XLSX.sheetnames(XLSX.readxlsx("myinputfile.xlsx"))
- **Import all data from a given sheet**: m = XLSX.readxlsx("myinputfile.xlsx")["Sheet1"][:]
- **Import all data from a specific interval**: m = XLSX.readxlsx("myinputfile.xlsx")["Sheet1"]["B2:D6"] or m = XLSX.readdata("myinputfile.xlsx", "Sheet1", "B2:D6")
- **Import data to a DataFrame**: df = DataFrame(XLSX.readtable("myinputfile.xlsx", "Sheet1"))

If you want to import to a DataFrame only a range of the sheet, you can specify in readtable() the columns you want to import as the third positional argument (e.g., B:D) and the first_row from which to import. However, at that point, you are probably better off using a package such as ExcelFiles, which allows directly importing into the DataFrame format.

To import from an OpenDocument Spreadsheet (ODS) format, you can use the OdsIO.jl (https://github.com/sylvaticus/OdsIO.jl) package; for example:

df = ods_read("myinputfile.ods"; sheetName="Sheet1", range=((2,2),(6,4)), retType="DataFrame")

## 5.2.4 Importing Data from JSON

JavaScript Object Notation (JSON) is a common open standard, language-independent data format that uses human-readable text (made of attribute–value pairs) to serialize and transmit data objects.

CHAPTER 5  INPUT/OUTPUT

Here I present the JSON3.jl (https://github.com/quinnj/JSON3.jl) package. Other common packages to read or export data in JSON format are JSON.jl (https://github.com/JuliaIO/JSON.jl) and Serde.jl (https://github.com/bhftbootcamp/Serde.jl). The latter one in particular is multiformat, as it supports serialization and deserialization in a multitude of formats (currently, JSON, TOML, XML, YAML, CSV, and Query). For YAML you also have the specific YAML.jl (https://github.com/JuliaData/YAML.jl) package, and for serialization in binary format, implementing the HDF5 specifications, you can use the JLD2.jl (https://github.com/JuliaIO/JLD2.jl) package.

Let's assume you have the following JSON data available in Julia as strings, representing a stand of trees from the Nottingham Forest:

---

```
json_string="""
{
    "lat": 53.204672,
    "long": -1.072370,
    "sp": "Oak",
    "trees": [
                {
                    "vol": 23.54,
                    "id": 1
                },
                {
                    "vol": 12.25,
                    "id": 2
                }
            ]
}
"""
```

---

105

There are two ways to import data in your program. The first one is to use JSON3.read(json_string, T); that is, to import the data as a specific type (that must be already defined):

```
struct ForestStand
    sp::String
    lat::Float64
    long::Float64
    trees::Array{Dict{String,Float64},1}
end
nott_for = JSON3.read(json_string, ForestStand)
```

nott_for will then be a ForestStand object.

Note that the order of the fields defined in the structure and in the imported JSON data don't need to match. What matters is the name and types of the fields (it was the opposite in previous versions).

The second way to import JSON data is to not specify any type for the imported object, as follows:

nott_for2 = JSON3.read(json_string)

nott_for2 will then be a JSON3.Object, in many way similar to a NamedTuple whose fields can be accessed with the normal dot notation; for example, nott_for2.trees[1].vol.

## 5.2.5 Accessing Web Resources

Julia offers several ways to access web resources, where again the most basic functionalities are provided in Standard Library packages and the more specialized functionalities are provided in third-party packages.

CHAPTER 5  INPUT/OUTPUT

In the following example, we set data_url ="https://archive.ics.uci.edu/ml/machine-learning-databases/housing/housing.data", which is the "house prices" dataset for Boston widely used as an example in several machine learning (ML) tutorials.

To save the dataset into a local file, you can use the download function from the Standard Library package Downloads.jl:

Downloads.download(data_url,"data.csv")

Alternatively, you can access web resources without necessarily saving them on disk by using the package HTTP.jl (https://github.com/JuliaWeb/HTTP.jl), which is quite large and also includes a server. Here, you just see the basic way to access (read) a web resource.

First, you obtain an HTTP.Message.Response using HTTP.get:

res         = HTTP.get(data_url)

Then you "extract" the actual data, in the form of a Vector{Uint8}, taking the body of the response:

data_binary   = res.body

If you need it, you can have a String representation of the data by calling the String constructor over this vector:

data_string = String(res.body)

---

### 🛈 Warning !

Calling the String constructor on a vector of Uint8 will empty the vector itself. If you still need the vector, use data_string = String(copy(res.body)) instead.

---

107

Finally, you can import the data into (for example) a DataFrame using the CSV.jl package:

```
data = CSV.read(data_binary, DataFrame, delim=" ",
ignorerepeated=true, header=false)
```

You can write the preceding code in a single passage using the convenient Pipe.jl (https://github.com/oxinabox/Pipe.jl) package (described in more detail in Chapter 10). In the following example, we import another ML dataset and clean it up on-the-fly (the original dataset has mixed field delimiters):

---

```
url_data = "https://archive.ics.uci.edu/ml/machine-learning-
databases/auto-mpg/auto-mpg.data-original"
data = @pipe HTTP.get(url_data).body                        |>
       replace!(_, UInt8('\t') => UInt8(' '))               |>
       CSV.File(_, delim=' ', missingstring="NA",
       ignorerepeated=true, header=false)                   |>
       DataFrame;
```

---

## 5.3 Writing (Output)

### 5.3.1 Writing to the Terminal

To write to the terminal, you can use the write(IO,T) function or print(IO,T), using the built-in stdout (default for print) or stderr streams:

- write(stdout, "Hello World");: Print the "Hello World" string to the terminal. Note the semicolon (;) at the end. Without it, the write method would also print the size (of characters) of the string printed.

CHAPTER 5   INPUT/OUTPUT

- `print("Hello World")`: Print the "Hello World" string to the terminal, without adding a newline at the end.
- `println("Hello World")`: Call `print("Hello World")` and add a newline.

---

> ℹ️ **Note**
>
> `write()` **vs.** `print()`
>
> The main difference between the two functions is that while `write()` outputs the raw bytes of the object, `print()` outputs a text representation of it. This representation is already defined for all the built-in types, and a default is provided for custom types (in contrast with `write()`, which would throw a `MethodError`).

---

The default textual representation of custom types can be modified by extending the `Base.show` function for the custom type, as shown in this example:

```julia
struct ACustomType
    x::Int64
    y::Float64
    z::String
end
foo = ACustomType(1,2,"MyObj") # Output: ACustomType(1, 2.0, "MyObj")
println(foo) # Output: ACustomType(1, 2.0, "MyObj")
write(stdout,foo) # Output: MethodError
import Base.show # Important ! Required to extend the `Base.show` method
function show(io::IO, ::MIME"text/plain", m::ACustomType)
```

```
    print(io,"A brief custom representation of ACustomType")
end
function show(io::IO, m::ACustomType)
    println(io,"An extended custom representation of
    ACustomType")
    println("($(m.x) , $(m.y)): $(m.z))")
end
foo # Output: A brief custom representation of ACustomType
println(foo) # Output: An extended custom representation of
ACustomType\n (1 , 2.0): MyObj)
write(stdout, foo) # Output: still a MethodError
```

## 5.3.2 Writing to a File

Writing to a file is similar to reading a file, but with the modality w (overwrite the file) or a (append to the existing data) instead of r in the open() function call. You must also, of course, use an output function as write or print over the IO object:

```
open("afile.txt", "w") do f   # "w" for writing
  write(f, "First line\n")    # \n for newline
  println(f, "Second line")   # Newline automatically added
                                by println
end
```

If you prefer to use the pattern "open > work on file > close" instead of using the do block, be aware that the operations will be flushed on the file on disk only when the IOStream is closed or when a certain buffer limit is reached.

Finally, note that write provides a convenient method where the first argument is a string representing a path. You can use it to shorten the preceding code, for example, as follows:

write("afile.txt", "First line\nSecond line")

## 5.3.3 Exporting to CSV

As the CSV format is based on simple text files, you could simply loop over the rows/columns of your data and manually "write" the data and delimiters into the file.

Alternatively, you can use either the DelimitedFiles.writedlm function or the CSV.write function, which do that for you:

```
DelimitedFiles.writedlm("myOutputFile.csv", my_matrix,";")
CSV.write("myOutputFile.csv", my_data, delim=';', decimal='.', missingstring="NA")
```

While this works out of the box when the data to export is a DataFrame, when the data is instead a standard matrix (Array{T,2}), you need first to transform it in a so-called MatrixTable using the function CSV.Tables.table(my_matrix):

CSV.write("myOutputFile.csv", Tables.table(my_matrix), delim=';', decimal='.', missingstring="NA", header=["field1", "field2","field3"])

write supports the following options: bufsize, delim, quotechar, openquotechar, escapechar, missingstring, dateformat, append, compress, writeheader, heade, newlin, quotestrings, decimal, transform, bom, partition.

## 5.3.4 Exporting to Excel and OpenDocument Spreadsheet (ODS) Files

Writing to Excel follows the common pattern of opening the file and operating on it within a do block.

```
XLSX.openxlsx("myExcelFile.xlsx", mode="w") do xf # w to write (or
                                                       overwrite)
                                                       the file
    sheet1 = xf[1]   # One sheet is created by default
    XLSX.rename!(sheet1, "new sheet 1")
    sheet2 = XLSX.addsheet!(xf, "new sheet 2") # We can add
                                                 further sheets
                                                 if needed

    sheet1["A1"] = "Hello world!"
    sheet2["B2"] = [ 1 2 3 ; 4 5 6 ; 7 8 9 ] # Top-right cell
                                               to anchor
                                               the matrix
end
```

This code creates a new file (if a file with the same name already exists, it will be overwritten). In order to "append" to an existing file without rewriting it, use mode="rw" instead:

```
XLSX.openxlsx("myExcelFile.xlsx", mode="rw") do xf
# rw to append to an existing file instead
    sheet1 = xf[1]   # One sheet is created by default
    sheet2 = xf[2]
    sheet3 = XLSX.addsheet!(xf, "new sheet 3") # We can add
                                                 further sheets
                                                 if needed
```

```
    sheet1["A2"] = "Hello world again!"
    sheet3["B2"] = [ 10 20 30 ; 40 50 60 ; 70 80 90 ]
    # Top-right cell to anchor the matrix
end
```

Individual cells can accept Missing, Bool, Float64, Int64, Date, DateTime, Time, or String values.

As shown in the last line of the preceding example, arrays (both Array{T,1} and Array{T,2}) are automatically broadcast to individual cells. This is not true for other data structures, like DataFrames. In the same way as XLSX.jl offers a specific function to facilitate importing to a DataFrame, it also offers writetable, a function to facilitate its exports. You can use it as XLSX.writetable("myNewExcelFile.xlsx", my_df) or you can pass to it a tuple (Array{Array{T,1},1}, Array{String}), that is columns of data and then field names. This format is easily obtainable from a DataFrame:

```
XLSX.writetable("myNewExcelFile.xlsx", sheet1=( [
[1, 2, 3], [4,5,6], [7,8,9]], ["f1","f2","f3"] ),
sheet2=(collect(DataFrames.eachcol(myDf)), DataFrames.
names(myDf) ))
```

Note that multiple sheets can be written at once, but the destination file must be new (or an error will be raised).

To export data to the OpenDocument Spreadsheet format, use ods_write(filename,data), where data is a dictionary in which the keys correspond to the locations of the top-left cells in the spreadsheet to which to export the data (a tuple of sheet name or position, row index, and column index) and the values are the actual data (a tabular structure such as Matrix, DataFrame, or even a Dictionary). For example:

```
ods_write("TestSpreadsheet.ods",Dict(
    ("TestSheet",3,2)=>[[1,2,3,4,5] [6,7,8,9,10]],
    ("TestSheet2",1,1)=>["aaa";;],
))
```

### 5.3.5 Exporting to JSON

Using the JSON3.jl (https://github.com/quinnj/JSON3.jl) package again and reusing the nott_for object defined in the "Importing Data from JSON" section, you can export the object to JSON simply by using json_string = JSON3.write(nott_for).

The returned string is a valid JSON string, but it's written as a single line and quoted. To display a human-readable version of it, use the @pretty macro provided by the same package: JSON3.@pretty json_string.

## 5.4 Other Specialized IO

All major data-storage formats have at least one Julia package to allow interaction with them. Here is a partial list (with my suggestions listed first):

- **XML**: EzXML.jl (https://github.com/bicycle1885/EzXML.jl) and LightXML.jl (https://github.com/JuliaIO/LightXML.jl)

- **HTML (web scraping)**: Gumbo.jl (https://github.com/JuliaWeb/Gumbo.jl) and Cascadia.jl (https://github.com/Algocircle/Cascadia.jl)

- **HDF5**: HDF5.jl (https://github.com/JuliaIO/HDF5.jl)

- **ARFF**: ARFFFiles.jl (https://github.com/cjdoris/ARFFFiles.jl)

# CHAPTER 6

# Metaprogramming and Macros

*Metaprogramming* is the technique that programmers use to write code (computer instructions) that, instead of being directly evaluated and executed, produces different code that is in turn evaluated and run by the machine.

For example, the following C++ macro is, in a broad sense, metaprogramming:

```
#define LOOPS(a,b,c)
  for (uint i=0; i<a; i++){
    for (uint j=0;j<b; j++){
      for (uint z=0;z<c; z++){
```

Here, the programmer, instead of writing the three for loops (assume that the program requires them in many places), can just write LOOPS(1,2,3) and this will be expanded and substituted for the loops in the actual code before it's run.

In this case, as in most programming languages, the "metaprogramming" usage remains fairly basic. First, it requires a different building step than the main one (in C++, this is done by the preprocessor). Second, and perhaps most important, the substitution is done at the level of a textual "search and replace," without any syntactical interpretation of the code as computer instructions.

CHAPTER 6  METAPROGRAMMING AND MACROS

In Julia and a few other languages (notably those derived or inspired by the Lisp programming language), the code written by the programmer is parsed and translated to computer instructions in what is known as an *abstract syntax tree (AST)*. At this point, the code is in the form of a data structure that can be traversed, created, and manipulated from within the language (without the need for a third-party tool or a separate building step). That means that the "code substitution" (later called a *macro*) can now be much more expressive and powerful, as you can directly manipulate computer instructions instead of text.

Aside from simplifying the code and avoiding tedious repetitions, metaprogramming and macros allow the implementation of domain-specific languages (DSLs), which enable package authors to create custom syntax tailored to specific problems, making the usage of their package more intuitive, readable, and concise for those familiar with the domain. You'll see an example of macros extensively used to create a domain-specific language when we discuss the JuMP algebraic modeling languages in Chapter 10.

This chapter introduces the concepts of *symbols* and *expressions*, and then explains how macros work.

## 6.1  Symbols

Symbols are intimately bound with the ability of Julia to represent the language's code as a data structure in the language itself. A *symbol* is a way to refer to a data object and still keep it in an unevaluated form. For example, when dealing with a variable, you can use a symbol to refer to the actual identifiers and not to the variable's value. Symbols can also refer to operators and any other parts of the (parsed) computer instructions. For example, `:myVar`, `:+`, and `:call` are all valid symbols.

To form a symbol, use the colon (`:`) prefix operator or the `Symbol()` function. For example:

   `a = :foo10` is equal to `a=Symbol("foo10")`

The Symbol function can optionally concatenate its arguments to form the symbol, as follows:

a = Symbol("foo",10)

To convert a symbol back to a string, use string(mysymbol).

## 6.2 Expressions

*Expressions* are unevaluated computer instructions. In Julia, even expressions are objects. Specifically, they are instances of the Expr type, whose fields are head, defining the kind of expression, and args, defining the array of elements. These can be symbols, primitive (not modifiable) values such as strings and numbers, or other subexpressions (from which the tree data structure is obtained).

Being objects, expressions are first-class citizens in Julia. They support all the operations generally available to other entities, such as being passed as an argument, returned from a function, modified, or assigned to a variable.

Expressions are normally created by parsing the computer code and in a second step they are evaluated. For example, the string "b = -(a+1)" when parsed becomes an expression. In the AST, this is shown as follows:

```
julia> expr = Meta.parse("b = -(a+1) # This is a comment");
julia> typeof(expr)
Expr
julia> dump(expr)
Expr
  head: Symbol =
  args: Array{Any}((2,))
    1: Symbol b
    2: Expr
      head: Symbol call
```

```
    args: Array{Any}((2,))
      1: Symbol -
      2: Expr
        head: Symbol call
        args: Array{Any}((3,))
          1: Symbol +
          2: Symbol a
          3: Int64 1
```

In the first line of this script, you start creating an expression object by parsing the string `"b=-(a+1) \# This is a comment"` and saving it in the expr variable. You then "dump" it in order to examine its internals. Note that comments have been stripped out and that this simple expression is already forming a tree. The equal sign is the head node, and the two arguments are the two child nodes: the variable b and a lower-level expression, which itself is a call to the unary - operator using as single parameter the expression corresponding to a+1. This shows you that, internally, operators are just shortcuts to normal functions taking parameters. For example, the plus operator in b = a+1 is actually a two-parameter function, b = +(a,1).

## 6.2.1 Expressions Definition

There are many ways to define an expression. The following four methods are all equivalent.

### 6.2.1.1 Parse a String

The first way to create an expression is what you just saw in the previous section: by parsing a string with `Meta.parse()`. For example:

`expr = Meta.parse("b = a+1")`

This is what Julia uses when parsing a .jl script or the REPL input.

## 6.2.1.2 Colon Prefix Operator

Expressions can be created using the same : prefix operator you saw for individual symbols, this time applied to a whole expression (given in parentheses):

```
expr = :(a+1)
```

## 6.2.1.3 Quote Block

An alternative to the :([...]) operator is to use the quote [...] end block:

```
expr = quote
         b = a+1
       end
```

## 6.2.1.4 Use the Exp Constructor with a Tree

The expression also can be directly constructed from a given tree by using the Exp constructor:

```
expr = Expr(:(=), :b, Expr(:call,:+,:a,1))
```

## 6.2.2 Evaluate Symbols and Expressions

To evaluate an expression or a symbol, use the eval() function:

```
expr = Meta.parse("3+2")
eval(expr) # 5
```

CHAPTER 6   METAPROGRAMMING AND MACROS

> **❶ Important !**
>
> The evaluation happens at the global scope, even if the eval() call is done within a function. That is, the expression being evaluated will have access to the global variables but not to the local ones:

```
a = 1
function foo()
  local a = 2
  expr = :(a + 1)
  return a+1, eval(expr)
end
foo() # out: (3,2)
```

This code is trivial, as it only contains literals—immutable values known at compile time. But what happens when the expression contains variables? Will the computation consider the value bound to the variable at the time the expression is composed or when it is evaluated? In Julia, you can choose this when you define the expression, as shown in the following snippet:

```
a = 1
expr1 = Meta.parse("$a + b + 3")
expr2 = :($a + b + 3)            # Equiv. to expr1
expr3 = quote $a + b + 3 end     # Equiv. to expr1
expr4 = Expr(:call, :+, a, :b, 3) # Equiv. to expr1
eval(expr1)                       # UndefVarError: b not defined
b = 10
```

120

```
eval(expr1)                     # 14
a = 100
eval(expr1)                     # Still 14
b = 100
eval(expr1)                     # 104
```

In the first line, you define the variable a and assign it a starting value, (i.e., you bound it to a certain object). In expressions 1 to 3, you use the dollar $ operator to interpolate ("unquote") the variable a. This indicates that you want to store in the expression the object to which a is bounded, and not the variable itself. This is why a has to exist beforehand.

Conversely, with b, you indicate that it is the identifier b and not its value that has to be stored in the expression. This variable doesn't even need to be defined at this point. Note that when you use the Expr constructor method to define expr4, you deal directly with unquoted entities. When you indicate a, you hence mean *the value bounded to it*, and to indicate an identifier instead, you need to use a symbol (such as :b).

When you try to evaluate the expression, at this point Julia tries to look up the b symbol and, not finding it, throws an error. You must define and assign a value to b before the expression can be evaluated.

Finally, note that whatever happens to the variable a after the definition of the expression has no effect on the evaluations of the expression, but modifying the value of b still affects it.

## 6.3 Macros

The heart of macros' power is the capability to represent code as expressions. Macros in Julia take one or more input expressions (and optionally literals and symbols), return modified expressions at parse time, and evaluate such expressions at runtime. By contrast, normal functions, at runtime, take the input values (the function arguments) and return a computed value.

Macros move the computation from the runtime to the compile time, as the expression is generated and compiled directly rather than requiring a runtime eval call like a function would.

## 6.3.1 Macro Definition

Defining a macro is similar to defining a normal function. The differences are that you use the macro keyword instead of the function keyword, and that in the body of the macro you assemble the expression that the macro will return—for example, by using a quote block in which the variables used as parameters are preceded by the dollar symbol. That makes sense, as the parameters of the macro are expressions, and so using the $ symbol for the variable bound to the parameters means injecting their content (an expression) within the broader context of the whole expression defined in the quote block.

Let's look at an example:

```
macro customLoop(controlExpr,workExpr)
  return quote
    for i in $controlExpr
      $workExpr
    end
  end
end
```

Here you use a macro to define a generic loop, where the generality is given by both the control you apply to the for loop and the expression(s) to be evaluated inside the loop.

## 6.3.2 Macro Invocation

After you define a macro, you can call it by using its name prefixed with the "at" symbol (@). The macro is then followed with the expressions that it accepts in the same row, separated by spaces (use the `begin` block syntax to group expressions together). For example:

```
a = 5
@customLoop 1:4 println(i)
@customLoop 1:a println(i)
@customLoop 1:a if i > 3 println(i) end
@customLoop ["apple", "orange", "banana"] println(i)
@customLoop ["apple", "orange", "banana"] begin print("i: "); println(i) end
```

You can see how the "expanded" macro is composed using another macro, called @macroexpand:

```
julia> @macroexpand @customLoop 1:4 println(i)
quote
    #= /path/to/source/file/metaprogramming.jl:65 =#
    for #92#i = 1:4
        #= /path/to/source/file/metaprogramming.jl:66 =#
        (Main.println)(#92#i)
    end
end
```

Note that a macro doesn't create a new scope, and variables declared or assigned within the macro may collide with variables in the scope at the point where the macro is actually called.

## 6.3.3 String Macros

Finally, a convenient type of macro (whose long but technically correct name is "nonstandard string literals") allows developers to perform compile-time custom operations on text entered as xxx"...text...". This is a custom prefix that's immediately attached to the text to be processed and entered as a string (triple quotes can be used as well, such as xxx"""...multi-line text...""").

For example, the following macro defines a custom eight-column display of the text:

```
macro print8_str(mystr)
  limits = collect(1:8:length(mystr))
  for (i,j) in enumerate(limits)
    st = j
    en = i==length(limits) ? length(mystr) : j+7
    println(mystr[st:en])
  end
end
```

String macros, whose name must end with _str, can then be called using a nonstandard string literal, where the prefix matches the macro name without the _str part:

```
julia> print8"123456789012345678"
12345678
90123456
78
```

You will see applications of these macros in the next chapter, where they will be used to process text representing code in a "foreign" language (C++, Python, or R).

# CHAPTER 7

# Interfacing Julia with Other Languages

The following third-party packages are covered in this chapter:

| | | |
|---|---|---|
| CxxWrap.jl | https://github.com/JuliaInterop/CxxWrap.jl | v0.16.0 |
| PythonCall.jl and JuliaCall (Python) | https://github.com/JuliaPy/PythonCall.jl | v0.9.23 |
| RCall.jl | https://github.com/JuliaInterop/RCall.jl | v0.14.4 |
| JuliaCall (R) | https://github.com/Non-Contradiction/JuliaCall | v0.16.5 |

Julia is a relatively new language (the first public release is dated February 2012), which means there is obvious concern regarding the availability of packages that implement its functionalities.

To address this concern, one specific area of Julia development addresses its capacity to interface with code written in other languages, and in particular to use the huge number of libraries available in those languages.

At its core, this has resulted in the capacity of the Julia language to natively call C and FORTRAN libraries. This in turn has allowed Julia developers to write packages to interface with the most common programming languages. I will discuss some of them in this chapter. Finally, this has allowed higher-level packages to leverage these "language-interface" packages to easily wrap existing libraries, written in other languages, and present an interface to them for Julia programmers.

To sum up, if you need a common functionality, it is highly likely that you will find a package that either implements the functionality directly in Julia (such as DataFrames.jl (https://github.com/JuliaData/DataFrames.jl)) or wraps an existing library in another language (for the same functionality, such as pandas.jl (https://github.com/JuliaPy/pandas.jl)).

I will discuss some of these packages in the second part of the book. The rest of this chapter will show you how to explicitly link Julia with other languages, either by calling functions implemented in other languages or by calling Julia functions from other languages.

> **ⓘ Note**
>
> When it could be ambiguous (such as when using prompts of different languages), I include the prompt symbols in the code snippets.

## 7.1 Julia  C

As stated, calling C code is native to the language and doesn't require any third-party code.

Let's first build a C library. I show this in Linux using the GCC compiler. The procedure in other environments is similar but not necessarily identical.

myclib.h:

```
extern int get2 ();
extern double sumMyArgs (float i, float j);
```

myclib.c:

```c
int get2 (){
 return 2;
}
double sumMyArgs (float i, float j){
 return i+j;
}
```

Note that you need to define the function you want to use in Julia as extern in the C header file.

You can then compile the shared library with gcc, a C compiler:

```
gcc -o myclib.o -c myclib.c
gcc -shared -o libmyclib.so myclib.o -lm -fPIC
```

You are now ready to use the C library in Julia:

```julia
const myclib = joinpath(@__DIR__, "libmyclib.so")
a = ccall((:get2,myclib), Int32, ())
b = ccall((:sumMyArgs,myclib), Float64, (Float32,Float32), 2.5, 1.5)
```

The ccall function takes as the *first* argument a tuple (function name, library path), where the library path must be expressed in terms of a full path, unless the library is in the search path of the OS. If it isn't, and you still want to express it relative to the file you are using, you can use the macro @__DIR__, which expands to the absolute path of the directory of the file itself. The variable hosting the full path of the library must be set constant, as the tuple acting as the first argument of ccall must be *literal*.

The *second* argument of ccall is the return type of the function. Note that while C int maps to either Julia Int32 or Int64, C float maps to Julia Float32 and C double maps to Julia Float64. In this example you could have used instead the corresponding Julia type aliases Cint, Cfloat, and Cdouble (within others) in order to avoid memorizing the mapping.

The *third* argument is a tuple of the types of the arguments expected by the C function. If there is only one argument, it must still be expressed as a tuple, such as (Float64,).

Finally, the remaining arguments are the arguments to pass to the C function.

We have just scratched the surface here. Linking C (or Fortran) code can become pretty complex in real-world situations, and consulting the official documentation (https://docs.julialang.org/en/v1/manual/calling-c-and-fortran-code/) is indispensable.

## 7.2 Julia ⇄ C++

As with C, the C++ workflow is partly environment dependent. This section demonstrates using CxxWrap.jl (https://github.com/JuliaInterop/CxxWrap.jl) on Linux for very simple functions that you compile manually with g++, import, and finally call from Julia.

For wrapping real C++ applications, you will probably want to use a C++ build chain or, on Windows, VS Code. Doing so is beyond the scope of this book, but in the CxxWrap repository you will find many examples that use CMake, as the project provides a regular CMake library. Instructions are also given on how to compile the C++ component on Windows using VS Code.

Whatever build system and environment you are using, the philosophy is the same: define the interface in C++ and then add a light wrapper in the form of Julia code.

## 7.2.1 A "Hello World" Example

This example shows you how to write a function that simply prints the classic "Hello world." Write the C++ source code and define the `cpp_hello` function as follows:

```
write("libcpp.cpp",
"""
#include <iostream>
#include "jlcxx/jlcxx.hpp"

void cpp_hello() {
  std::cout << "Hello world from a C++ function" << std::endl;
  return;
}

JLCXX_MODULE define_julia_module(jlcxx::Module& mod) {
  mod.method("cpp_hello", &cpp_hello);
}
""")
```

As you can see, in the C++ source, it is enough to add the C++ headers of CxxWrap and a small section to register the function for CxxWrap.

As shown next, store a few paths with the C++ headers you need to compile. Again, for large libraries you will probably want to use a CMakeLists.txt file instead.

```
using CxxWrap
cxx_include_path   = joinpath(CxxWrap.prefix_path(),"include")
julia_include_path = joinpath(Sys.BINDIR,"..","include","julia")
```

You can now compile the C++ source into a shared library:

```
cmd = `g++ -shared -fPIC -o libcpp.so -I $julia_include_path
-I $cxx_include_path  libcpp.cpp`
run(cmd)
```

On top of the CxxWrap includes you need to add the Julia ones, as CxxWrap refers to them. Back in Julia, you simply need to load the library and call the function:

```
@wrapmodule(() -> joinpath(pwd(),"libcpp"))
cpp_hello()
```

## 7.2.2 Passing Arguments and Retrieving Data

The example in the previous section doesn't have any data exchange between the caller (Julia) and the called function (in C++).

It ends up that most primitive types are automatically converted by CxxWrap. In the following example, the C++ function takes two integers and returns a floating point. As you can see, there isn't anything that the developer has to account for:

```
write("libcpp.cpp",
"""
#include "jlcxx/jlcxx.hpp"

double cpp_average(int a, int b) {
  return (double) (a+b)/2;
}
```

```
JLCXX_MODULE define_julia_module(jlcxx::Module& mod) {
  mod.method("cpp_average", &cpp_average);
}
""")
```

The C++ code can be compiled, loaded, and invoked as before, with the important caveat that it is not possible to register the same library twice with the @wrapmodule, so either it must be renamed or Julia must be restarted.

## 7.2.3 Functions with STD Classes

Functions that work with—or return—C++ Standard Library (STD) objects can still be interfaced, but you need to take a few precautions. First, you need to add the specific CxxWrap headers. Second, since CxxWrap uses modern C++ code, you need to compile the library with the --std=c++20 flag. Finally, you may need to manually convert the function argument or returned object.

Let's see how to do this with a couple of functions. The first one, cpp_sum, takes an STD vector and returns an STD string. The second, cpp_multiple_averages, goes a bit deeper and works on nested STD structures:

```
write("libcpp.cpp",
"""
#include <string>
#include <iostream>
#include <vector>

#include "jlcxx/jlcxx.hpp"
#include "jlcxx/functions.hpp"
#include "jlcxx/stl.hpp"
```

```
using namespace std;

string cpp_sum (std::vector< double > data) {
  double total = 0.0;
  double nelements = data.size();
  for (int i = 0; i< nelements; i++){
    total += data[i];
  }
  std::stringstream ss;
  ss << "The sum is " << total << endl;
  return ss.str();
}

std::vector<double> cpp_multiple_averages (std::vector<
std::vector<double> > data) {
  std::vector <double> averages;
  for (int i = 0; i < data.size(); i++){
    double isum = 0.0;
    double ni= data[i].size();
    for (int j = 0; j< data[i].size(); j++){
      isum += data[i][j];
    }
    averages.push_back(isum/ni);
  }
  return averages;
}

JLCXX_MODULE define_julia_module(jlcxx::Module& mod) {
  mod.method("cpp_sum", &cpp_sum);
  mod.method("cpp_multiple_averages", &cpp_multiple_averages);
}
""")
```
--------------------------------------------------------------

You compile it and wrap it as follows:

```
cmd = `g++ --std=c++20 -shared -fPIC -o libcpp.so -I $julia_include_path -I $cxx_include_path  libcpp.cpp`
run(cmd)
@wrapmodule(() -> joinpath(pwd(),"libcpp"))
```

Suppose you have the following data:

```
data_julia = [1.5,2.0,2.5]
```

If you were to try to call `cpp_sum(data_julia)` directly, you would get a `MethodError`. Instead, you need to wrap the data into a `StdVector`, a type provided by `CxxWrap`:

```
data_sum    = cpp_sum(StdVector(data_julia))
```

The returned value is also a specific `CxxWrap` object, but it is a child of `AbstractString`, and in practice it can be used as a standard Julia string.

The same applies to the `cpp_multiple_averages` function:

```
data_julia = [[1.5,2.0,2.5],[3.5,4.0,4.5]]
data       = StdVector(StdVector.(data_julia))
data_avgs  = cpp_multiple_averages(data) # [2.0, 4.0]
```

Because the function argument is a nested vector of a vector, you also need to nest the `CxxWrap StdVector` constructor.

This section only scratches the surface of the possibilities of interfacing C++ and Julia with `CxxWrap`. Apart from the building steps, there are several examples in its repository, such as how to wrap C++ classes and how to deal with their inheritance.

## 7.3 Julia ⇌ Python

To call Python code in Julia or, conversely, call Julia code in Python, you can use PythonCall.jl and JuliaCall (https://github.com/JuliaPy/PythonCall.jl) respectively, where PythonCall.jl refers to the name of the Julia package to call Python code, and JuliaCall is the name of the Python package to call Julia code. They are hosted on the same repository, as they share most of the same code.

Some of their features include the following:

- They can automatically download and install a local copy of Python (Julia), private to your Julia (Python) project, in order to avoid messing with version dependencies and providing a consistent environment within Linux, Windows, and macOS.
- They provide manual or automatic conversion between Julia and Python types.
- They are very simple to use.

> **! Warning !**
>
> Don't confuse PythonCall.jl with PyCall.jl (https://github.com/JuliaPy/PyCall.jl). PyCall.jl is an older Julia/Python package that differs mostly in the default way of making Python available to Julia. Similarly, do not confuse the JuliaCall package for Python (discussed here) with the JuliaCall package for R discussed in the next section, as they are unrelated (except for their objective).

CHAPTER 7  INTERFACING JULIA WITH OTHER LANGUAGES

## 7.3.1 PythonCall Installation

Concerning the first point in the previous list, `PythonCall` by default installs a Conda-based, project-specific version of Python (using `CondaPkg.jl` (https://github.com/JuliaPy/CondaPkg.jl)) on the first `using PythonCall` command.

If you want instead to use your existing Python interpreter, you can instruct `PythonCall` to use it by setting the environment variable `ENV["JULIA_CONDAPKG_BACKEND"] = "Null"` and, if python is not in your path, `ENV["JULIA_PYTHONCALL_EXE"] = "/path/to/python"` before using `PythonCall`. Note that as environment variables are session specific, you need to put them in your script; for example:

```julia
# Use the system default Python
using Pkg
cd(@__DIR__)
Pkg.activate(".")
ENV["JULIA_CONDAPKG_BACKEND"] = "Null"
ENV["JULIA_PYTHONCALL_EXE"]   = "/home/user/.pyenv/shims/python"
using PythonCall
```

The rest of this section assumes that you have stuck with the standard installation of `PythonCall`.

## 7.3.2 Evaluate Python Code in Julia

You evaluate Python code in a Julia program by using the `@pyexec (inputs) => python_code => output_in_julia` macro. For example:

CHAPTER 7   INTERFACING JULIA WITH OTHER LANGUAGES

```
i = 3
@pyexec (i=i, j=4) => """
a=i+j
b=i/j
""" => (a::Int64,b::Float64)
```

This code creates two Julia variables, a and b, and assigns them the values 7 and 0.75, respectively.

You can use the same macro to define Python functions that you can then call directly in Julia:

```
@pyexec """
def python_sum(i, j):
    return i+j
""" => python_sum
@pyexec """
def get_ith_element(n):
    a = [0,1,2,3,4,5,6,7,8,9]
    return a[n]
""" => get_ith_element
```

You can now call these functions:

```
c = @pipe python_sum(3,4)           |>
pyconvert(Int64,_)            # 7
d = @pipe python_sum([3,4],[5,6]) |> pyconvert(Vector{Int64},_)
# [8,10]
typeof(d)                                 # Vector{Int64}
e = @pipe get_ith_element(i)        |>
pyconvert(Int64,_)         # 3
```

136

Note that the conversions of the inputs is automatic, but you still need to convert the outputs. The result of the two functions is a Py object that needs to be converted using the function pyconvert(T,py_object).

Also note in the last line of the previous example that PythonCall doesn't attempt index conversion (Python arrays are zero-based, while Julia arrays are one-based): calling the Python get_ith_element() function with 1 as an argument will retrieve what *in Python* is the element "1" of the array, that is, the second element.

If your Python functions are defined in a script, you can execute the script with the help of the read function:

```
pyexec(read("python_code.py", String),Main)
@pyexec (i=i, j=4) => "f = python_sum(i,j)" => (f::Float64)
```

### 7.3.3 Use Python Packages in Julia

To install a Python dependency and make it available in a Julia project, use the CondaPkg.jl commands add_channel(channel) and add(pkg; version, channel) (suggested) or add_pip(pkg).

The following example shows how to install the Python ezodf module (https://github.com/T0ha/ezodf) to create an OpenDocument Spreadsheet (ODS), mimicking the code of the OdsIO.jl package introduced in Chapter 5:

```
Pkg.add("CondaPkg")      # only once, saved on Project.toml
import CondaPkg          # only once
CondaPkg.add("ezodf")    # only once, saved on CondaPkg.toml
```

By adding a Python package, CondaPkg will automatically create a CondaPkg.toml file in the current environment. This is very similar in structure and scope to the Julia Project.toml file, defining the packages used in the project. If you share your code with someone or use a Git repository, don't forget to include this file.

You are now ready to import and work with the ezodf module:

```julia
const ez = pyimport("ezodf")  # Equiv. of Python `import ezodf as ez`
dest_doc = ez.newdoc(doctype="ods", filename="an_ods_sheet.ods")
sheet = ez.Sheet("Sheet1", size=(10, 10))
dest_doc.sheets.append(sheet)
dcell1 = sheet[(2,3)] # This is cell "D3", not "B2"
dcell1.set_value("Hello")
sheet["A9"].set_value(10.5)
dest_doc.backup = false
dest_doc.save()
```

The usage of the module in Julia follows the Python API. The module is imported and assigned to a shorter alias, ez. You can then directly call its functions with the usual Python syntax, module.function().

The dest_doc object returned by newdoc is a generic Py object. You can access its attributes and methods with a_Py_object.attribute and a_Py_object.method(), respectively, and if the object represents a Python container, you can select a specific element by indexing it with a_Py_object[index_key].

Finally, note again that index conversion is *not* automatically implemented: when asking for sheet[(2,3)], the values are interpreted as *Python*-based indexes, and cell D3 of the spreadsheet is returned, not B2.

CHAPTER 7  INTERFACING JULIA WITH OTHER LANGUAGES

## 7.3.4 JuliaCall (Python Package) Installation

The reverse scenario of embedding Julia code in a Python script or terminal is equally of interest, as in many cases it provides substantial performance gains for Python programmers, and it may be easier than embedding C or C++ code.

You can install `JuliaCall` by using `pip` as follows:

```
$ python -m pip install --user juliacall
```

Next, you can obtain a reference to the `Main` module of Julia with

```
>>> from juliacall import Main as jl
```

At this point, `JuliaCall` assesses whether you have a valid Julia executable in your path. If so, it will reuse it; otherwise, it will automatically download and use one.

## 7.3.5 Evaluate Julia Code in Python

You can now evaluate Julia code using the `seval()` function:

```
>>> jl.seval("""
... function funny_prod(is, js)
...     prod = 0
...     for i in 1:is
...       for j in 1:js
...         prod += 1
...       end
...     end
...     return prod
... end
... """)
```

## CHAPTER 7   INTERFACING JULIA WITH OTHER LANGUAGES

You can then call the preceding function in Python as `jl.funny_prod(2,3)`. Alternatively, you can use `seval` also to call the function. This is particularly useful if you want to broadcast it—that is, apply the function to each element of a given array. This is obtained in Julia using dot notation, such as `jl.seval("funny_prod.([2,3],[4,5])")`.

Note that the objects returned from calling the Julia functions are Python objects. For a mutable object, this is a wrapper to the Julia object. In particular for arrays, this object supports the Numpy API and hence can be generally used as such. For functions that require "true" Numpy arrays, you can convert it to a Numpy array using `obj.to_numpy()`.

If the evaluated Julia code creates a new variable, you can access it in Python by using the reference to the `Main` Julia module. For example:

```
>>> jl.seval("i = [1,2,3]")
>>> jl.i
```

Alternatively, you may want to define all your functions in a Julia script and "include" it. Assume `julia_code.jl` is a file made of the following Julia code:

```julia
function hello_world()
    println("Hello World!")
end
function julia_sum(i, j)
  return i+j
end
function get_ith_element(n)
  a = [0,1,2,3,4,5,6,7,8,9]
  return a[n]
end
```

You can access its functions in Python with

```
>>>> jl.seval("include(\"julia_code.jl\")")
>>>> jl.hello_world() # Prints `Hello World!`
>>>> a = jl.julia_sum([1,2,3],[4,5,6])
>>>> b = jl.get_ith_element(1)  # Returns `0`, the "first"
                                       element for Julia
```

Note that you now get *the Julia way* of indexing (one-based).

## 7.3.6 Use Julia Packages in Python

To use Julia packages, you have two options. In both cases, which environment will be used to store the package information depends on the way you are running Python:

- If you are in a virtual or Conda Python environment, the Julia environment is put on the same directory as the Python environment.
- If you are using the global Python interpreter already present on your machine, the Julia default environment will be set to [USER_HOME_FOLDER]/.julia/environments/pyjuliapkg/.

The first method to add a Julia package is simply to evaluate the Julia code used to install packages; for example:

```
jl.seval("""
   using Pkg
   Pkg.add("DataFrames")
""")
```

The second method is to use `PyJuliaPkg`, a package to manage Julia packages (environments) specific to a certain Python project, and save the environment information in a `juliapkg.json` file, in a subfolder of the same Julia environment folder.

You install `JuliaPkg` with `pip`:

```
$ python -m pip install --user juliapkg
```

You then import `juliapkg` and add a package to the environment with `juliapkg.add(pkg, uuid)` and `juliapkg.resolve()`, as shown in this example:

```
import juliapkg as jpkg
jpkg.add("CSV", "336ed68f-0bac-5ca0-87d4-7b16caf5d00b")
# jpkg.add(pkg, uuid)
jpkg.resolve()
```

The UUID of the package to install is mandatory and can be found in the `Project.toml` on its Git repository.

> **⚠ Warning !**
>
> At the time of writing, the two methods just described don't work well together. In particular, if you use `juliapkg`, be aware that the `resolve()` step will simply copy the environment information from the `juliapkg.json` file to the `Project.toml` file, overriding whatever you may have already set there using the first method.

## 7.4 Julia ⇌ R

As with Python, with R you have the choice to either run R code (and access its huge set of specialized data science libraries) in Julia (with RCall.jl (https://github.com/JuliaInterop/RCall.jl)) or, conversely, execute efficient Julia code from within R (with JuliaCall (https://github.com/Non-Contradiction/JuliaCall)).

### 7.4.1 RCall.jl Installation

There are two ways to tell RCall.jl where to find the R version to use. The classical, simpler method is to use environmental variables at installation/build time. This is described next. Alternatively, you can specify the R version to use on each individual Julia project by using the Julia Preferences.jl (https://github.com/JuliaPackaging/Preferences.jl) package. Just add a LocalPreferences.toml file in the same directory as the project's Project.toml file with the following two entries:

---

**[RCall]**
Rhome = "/path/to/env/lib/R"
libR = "/path/to/env/lib/R/lib/libR.so"

---

You can find more information in the RCall documentation (https://juliainterop.github.io/RCall.jl/stable/installation/#Customizing-the-R-installation-using-Julia's-Preferences-system). The rest of this section assumes that you are installing RCall using the environmental variables method.

When you add the RCall package, it should automatically download and install R if it doesn't detect a local R installation (with R >= 3.4.0). If this doesn't work for any reason, you can always install R before installing RCall.

CHAPTER 7   INTERFACING JULIA WITH OTHER LANGUAGES

You can choose the version of R that RCall.jl will use by setting the environmental variable R_HOME to the base directory of your R installation (which you can retrieve by typing R.home() at your local R prompt). If you want to force RCall.jl to download a "private" Julia version of R, even if a local version is available, use instead ENV["R_HOME"]="*".

```
ENV["R_HOME"]="/path/to/the/R/base/directory" # E.g. /usr/lib/R/"
using Pkg
Pkg.add("RCall")
Pkg.build("RCall")
```

## 7.4.2 Evaluate R Code in Julia

You can access an embedded R prompt from Julia typing the dollar sign ($) symbol in the keyboard (and then go back to the Julia REPL with the Backspace key). Note that copy it would not work, it needs to be typed in the keyboard. You can then access (from the embedded R prompt) Julia symbols or expressions using $ again or using the @rput macro. Conversely, you can retrieve R variables back to Julia with the @rget macro:

```
julia> a = [1,2]; b = 3
julia> @rput a        # Explicitly transfer the `a` variable and
                      the object to which it is bonded to R
R>      c = a + $b    # Access a and, through the dollar
                      symbol, b, in R
julia> @rget c        # Explicitly transfer the `c` variable and
                      the object to which it is bonded to Julia
julia> c              # [4,5]
```

144

CHAPTER 7  INTERFACING JULIA WITH OTHER LANGUAGES

You don't need to convert the data, as RCall supports automatic conversions to and from the basic Julia types (including arrays and dictionaries), as well as popular tabular data formats, such as data frames.

As an alternative to using the embedded R terminal, you can embed R code in Julia. You both define and call the R functions with the R"..." string macro, and in the function call, you can directly use your Julia data:

```
using RCall
R"""
r_sum <- function(i,j) i+j
get_ith_element <- function(n) {
  a <- c(0,1,2,3,4,5,6,7,8,9)
  return(a[n])
}
"""
i = [3,4]
a = rcopy(R"r_sum"(3,4))          # 7
b = rcopy(R"r_sum"(i,[5,6]))      # [8,10]
b = rcopy(R"r_sum($i,c(5,6))")    # [8.0,10.0] (alternative)
c = rcopy(R"r_sum"([3,4],5))      # [8,9]
d = rcopy(R"get_ith_element"(1))  # 0.0
```

As with PythonCall, the results of calling the R functions defined with R"..." are not yet converted to exploitable Julia objects, but remain as RObjects. To convert the RObjects to standard Julia objects, you can use the rcopy function and, if you want to force a specific conversion, you can still rely on convert:

```
convert(Array{Float64}, R"r_sum"(i,[5,6]))   # [8.0,10.0]
convert(Array{Int64}, R"r_sum($i,c(5,6))")   # [8,10]
```

CHAPTER 7  INTERFACING JULIA WITH OTHER LANGUAGES

Finally, with R, you don't have to worry about the indexing convention, as R and Julia both adopt one-based array indexing.

### 7.4.3 Use R Packages in Julia

The installation of an R package depends on the way `RCall` has been instructed to work with R. If you set `RCall` to download the default Conda-based version of R, you need to use Conda also to install R packages:

```
import Conda
Conda.update()
Conda.add("r-ggplot2")
```

Note that if the `Conda.update` command also updates R, you may need to restart Julia to ensure that the R package works correctly with R.

If you instead used an already existing installation of R (not Conda based), you can use the `R"..."` string macro to install R packages:

```
using RCall
R"options(download.file.method='wget')" # Needed in some setups
R"install.packages('ggplot2', repos='http://cran.us.r-project.org')"
```

Either way, you can now import the packages in R. Although `RCall` provides two specialized macros for working with R libraries, `@rlibrary` and `@rimport`, these are inefficient, as data would be copied multiple times between R and Julia. Just use the `R"..."` string macro instead:

CHAPTER 7   INTERFACING JULIA WITH OTHER LANGUAGES

```
mydf = DataFrame(deposit = ["London","Paris","New-York","Hong-
Kong"]; q = [3,2,5,3] )  # Create a DataFrame ( a tabular
                          structure with named cols) in Julia
R"""
  library(ggplot2)
  ggplot($mydf, aes(x = q)) +
  geom_histogram(binwidth=1)
"""
```

**Note**   In macOS, you may need the XQuartz display server for X Window System (https://www.xquartz.org/) to display the chart.

## 7.4.4 JuliaCall (R Package) Installation

To go the reverse route and embed Julia code within an R workflow, you can use the R package `JuliaCall`:

```
> install.packages("JuliaCall")
```

You can then use Julia:

```
> library(JuliaCall)
> julia_setup()
```

147

Note that the `julia_setup` function needs to be called every time you start a new R session, not just when you install the `JuliaCall` package. Several situations may arise:

1. If you have julia and it's in your path, use `julia_setup()` as just shown.

2. If you have julia but it isn't in the path of your system, or you have multiple versions and you want to specify the one to work with, pass the `JULIA_HOME = "/path/to/julia/binary/executable/directory"` parameter (e.g., JULIA_HOME = "/home/myUser/lib/julia-1.11.0/bin") to the `julia_setup` call.

3. If you have julia on your path but you want to force R to work with a "private" version of Julia, use `install_julia()` before `julia_setup` to let R download and install a private copy of Julia (only once).

4. If you don't have Julia installed, pass the `installJulia = TRUE` parameter to the `julia_setup` call.

`JuliaCall` depends on some things (like object conversion between Julia and R) from the Julia `RCall` package. If you don't already have the package installed in Julia, Julia will try to install it automatically.

## 7.4.5 Evaluate Julia Code in R

As expected, `JuliaCall` offers multiple ways to access Julia in R.

CHAPTER 7  INTERFACING JULIA WITH OTHER LANGUAGES

You can embed Julia code directly in R using the julia_eval() function:

```
> funny_prod_r <- julia_eval('
+   function funny_prod(is, js)
+     prod = 0
+     for i in 1:is
+       for j in 1:js
+         prod += 1
+       end
+     end
+     return prod
+   end
+ ')
```

You can then call this function in R either as funny_prod_r(2,3), julia_eval("funny_prod(2,3)"), or julia_call("funny_prod",2,3).

Alternatively, you may have all your Julia functions in a file. You are going to reuse the julia_code.jl script you used in the Python JuliaCall package. You can access its functions in R with

```
> julia_source("julia_code.jl") # Include the file
> julia_eval("hello_world()") # Prints `Hello World!` and
                                returns NULL
> a <- julia_call("julia_sum",c(1,2,3),c(4,5,6)) # Returns
                                                  `[1] 5 7 9`
> as.integer(1) %>J% get_ith_element -> b   # Returns `0`,
                                              the "first"
                                              element for both
                                              Julia and R
```

149

The preceding example highlights the usage of the pipe operator that is very common in R (you will see an equivalent in Julia in Chapter 12). The %>J% syntax is a special "version" of it, provided by JuliaCall, allowing you to mix Julia functions in a left-to-right data transformation workflow.

While other "convenience" functions are provided by the package, julia_source, julia_eval, and julia_call should suffice to accomplish any task you may need in Julia.

## 7.4.6  Use Julia Packages in R

Julia packages can be installed by evaluating the corresponding Julia syntax to install packages. For example, to install the DataFrames package, use

```
> julia_eval('using Pkg; Pkg.add("DataFrames")') # only once
```

You can then use the package:

```
> julia_eval("using DataFrames")
> rdf <- julia_eval('DataFrame(A=["a","b","c"],B=[1,2,3])')
> str(rdf)
'data.frame':    3 obs. of  2 variables:
 $ A: chr  "a" "b" "c"
 $ B: int  1 2 3
```

As you can see, the Julia DataFrame object has been automatically converted to the R data.frame object.

# CHAPTER 8

# Efficiently Write Efficient Code

The following third-party packages are covered in this chapter:

| BenchmarkTools.jl | https://github.com/JuliaCI/BenchmarkTools.jl | v1.5.0 |
| JuliaInterpreter.jl | https://github.com/JuliaDebug/JuliaInterpreter.jl | v0.9.34 |
| Julia VS Code extension | https://www.julia-vscode.org | v1.105.2 |

The objective of this chapter is to present topics that, while maybe not strictly essential, are nevertheless very important for writing efficient code in an efficient manner—that is, code that runs fast and is easy (and fun) to write. This chapter first deals with code performance, highlighting the issues that may make your code run in a suboptimal way. It then discusses tools that enable you to check that the program is following the intended behavior, facilitating early bug determination. The chapter wraps up with the constructs that you can employ to isolate, and possibly correct, problems that arise at runtime.

CHAPTER 8   EFFICIENTLY WRITE EFFICIENT CODE

# 8.1 Performance

Julia is already relatively fast when working with data objects of type Any; that is, when the type of the object remains unspecified until execution. But when the compiler can infer a specific type (or a union of a few types), Julia programs can run with the same order of magnitude as C programs.

The good news is that programmers don't typically need to specify variables types. They are inferred directly by the compiler to produce efficient, specialized code. It is only in rare situations that the compiler can't infer a specific type.

In this section, you learn not only how to measure (or benchmark) Julia code and how to avoid the problem of uninferred types, but also a few other tips specific to Julia that can help improve performance.

## 8.1.1 Benchmarking

When you highlight code in your editor and run it to evaluate the corresponding instructions, Julia performs two separate tasks: it compiles the code (and eventually its dependencies) and then evaluates it.

---

> **❗ Important !**
>
> Unless you are interested in how effective Julia is for interactive development, you should benchmark and/or profile only the second operation (the evaluation), and not the code compilation. Hence, if you want to use the simplest benchmarking macro, `@time [a Julia expression]`, you should run it twice and discard the first reading, which includes compile time.

---

Let's consider for example a function that computes the Fibonacci sequence, where each number is the sum of the Fibonacci function applied to the previous two numbers (i.e., $F_n = F_{n-1} + F_{n-2}$), with $F_0$ and $F_1$ defined as 0 and 1, respectively.

A naive recursive implementation would look like this (note that much faster algorithms to compute Fibonacci numbers exist):

```julia
julia> function fib(n)
           if n == 0 return 0 end
           if n == 1 return 1 end
           return fib(n-1) + fib(n-2)
       end
fib (generic function with 1 method)
```

If you start a new Julia session, you could run this code and try to time it with the @time macro. You would obtain something similar to this:

```julia
julia> @time @eval fib(10)
  0.007413 seconds (1.15 k allocations: 75.969 KiB, 97.22% compilation time)
55
julia> @time @eval fib(10)
  0.000199 seconds (43 allocations: 2.109 KiB)
55
```

Although the @time macro can accept any valid Julia expression, it's most often used with function calls.

You can see that the first time and memory usage information reported includes the compilation of the function. You can also deduce that the compilation of a function happens "just in time" (JIT), which is the first time the function is called, and not when it is defined.

**Note**

In the preceding example, I added an @eval expression in the timing because in recent versions of Julia, @time has become "smarter," and for very simple functions such as the fib one, it performs a precompilation before the actual timing and hence the difference between the first and second timing would be lost.

Further, is the function time *really* 0.0002 seconds? That's only one example of a variable that includes some stochastic components related to the status of the machine. A better approach is to have some *average* timing, with a number of draws that makes the mean statistically significant.

The BenchmarkTools.jl (https://github.com/JuliaCI/BenchmarkTools.jl) package does exactly that: it provides @benchmark, an alternative macro that automatically skips the compilation time and runs the code several times in order to report an accurate timing.

Applied to this function, you would have the following:

```
julia> @benchmark fib(10)
BenchmarkTools.Trial: 10000 samples with 216 evaluations.
 Range (min … max):  344.231 ns …   2.392 µs  ┊ GC (min … max): 0.00% … 0.00%
 Time  (median):     344.431 ns               ┊ GC (median):    0.00%
 Time  (mean ± σ):   381.890 ns ± 140.819 ns  ┊ GC (mean ± σ):  0.00% ± 0.00%

  344 ns           Histogram: log(frequency) by time       795 ns <

 Memory estimate: 0 bytes, allocs estimate: 0.
```

This gives you a much more precise benchmark result.

## 8.1.2 Profiling

While benchmarking gives information about the *total* time spent running the given instructions, often you want to know how the time is allocated within the function calls, in order to find bottlenecks. This is the task of a *profiler*.

Julia comes with an integrated statistical profiler (in the `Profile` package of the standard library). When invoked with an expression, the profiler runs the expression and records, every $x$ milliseconds, the line of code where the program is at that exact moment. The more often a line of code is hit, the more computationally expensive it is.

Using this sampling method, at the cost of losing some precision, profiling can be very efficient. It will lead to a very small overhead compared to running the code without it. And it is very easy to use:

- `@profile myfunct()`: Profile a function (remember to run the function once before, or you will also measure the JIT compilation).
- `Profile.print()`: Print the profiling results.
- `Profile.clear()`: Clear the profile data.

Profile data is accumulated until you clear it. This allows developers to profile even very small (fast) functions, where a single profiling run would be inaccurate. Just profile them multiple times, such as:

`@profile (for i = 1:100; foo(); end)`

For example, assume that you have the following function to profile in a file called `myCode.jl`, with the numbers on the left being the line numbers:

```
33 function lAlgb(n)
34    a = rand(n,n) # matrix initialization with random numbers
35    b = a + a     # matrix sum
36    c = b * b     # matrix multiplication
37 end
```

## CHAPTER 8   EFFICIENTLY WRITE EFFICIENT CODE

If you run @profile (for i = 1:100; lAlgb(1000); end) (again, after having run the function once for the JIT compilation) and then run Profile.print(), you may end up with something like the following output (which has been partially cut for space reasons):

```
----------------------------------------------------------
Overhead ¦ [+additional indent] Count File:Line; Function
==========================================================
[...]
     ¦112     /full/path/to/myCode.jl:34; lAlgb(n::Int64)
     ¦ 112    ...stdlib/v1.10/Random/src/Random.jl:279; rand
     ¦   112  ...stdlib/v1.10/Random/src/Random.jl:291; rand
[...]
     ¦216     /full/path/to/myCode.jl:35; lAlgb(n::Int64)
     ¦ 216    @Base/arraymath.jl:16; +(A::Matrix{Float64}, Bs::Matrix{Float64})
     ¦   216  @Base/broadcast.jl:892; broadcast_preserving_zero_d
     ¦    216 @Base/broadcast.jl:903; materialize
[...]
     ¦307     /full/path/to/myCode.jl:36; lAlgb(n::Int64)
     ¦ 307    ...stdlib/v1.10/LinearAlgebra/src/matmul.jl:113; *(A::Matrix{Float64}, B::Matrix{Float64})
[...]
Total snapshots: 15174. Utilization: 13% across all threads and tasks.
Use the `groupby` kwarg to **break** down by thread and/or task.
----------------------------------------------------------
```

CHAPTER 8   EFFICIENTLY WRITE EFFICIENT CODE

The format of each line in the output is the number of samples in the corresponding line and in all downstream code, the file name, the line number, and then the function name. The important lines are those related to the individual function's operation. From the profile log, you see that the initialization of the matrix with a random number (row 34) has been hit 112 times, the matrix summation (row 35) 216 times, and the matrix multiplication (row 36) 307 times. This helps you to determine the *relative* time spent within the lAlgb function, with the matrix multiplication being the most computationally intensive operation.

In real use, things quickly become complicated, and you may appreciate a graphical representation of the call stack with the profiled data. To obtain that, you can use the profiler integrated in the Julia extension of VS Code. Just replace @profile [expression] with @profview [expression] while using VS Code.

As shown on the left in Figure 8-1, the VS Code profiler provides a nice flame chart of the call stack, with the width proportional to the hit count. The chart is clickable, so you can zoom in on the part of the stack you are interested in inspecting. Further, the source code is highlighted with semitransparent bars proportional to the hit count.

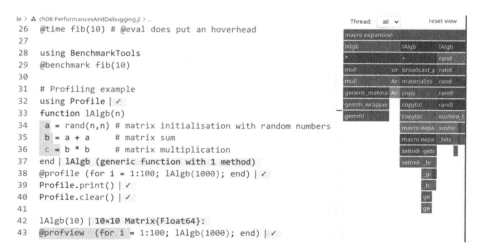

***Figure 8-1.*** *Profiling session in VS Code*

## 8.1.3 Type Stability

A function is said to be *type-stable* when its return type depends only on the types of the input, not on their values. While type-unstable functions have their return type compiled as Any, and are converted to a specific type only at runtime, type-stable functions can be optimized by the compiler, as it is able to compile specialized versions for all the possible types the function can be called with. Further, type stability is essential in retaining the capacity of the compiler to infer the types of the various objects across the various chained calls.

It is important to note that the problem of type instability is *not* that parameters aren't annotated with a specific type. Often the compiler will still be able to list, at compile time, all the possible types of the parameters for that function, and still produce performant code, even without explicit type annotation.

Let's look at an example of a type-unstable function:

```
function f_unstable(x)    # Type unstable
    out_vector = [1,2.0,"2"]
    if x < 0
        return out_vector[1]
    elseif x == 0
        return out_vector[2]
    else
        return out_vector[3]
    end
end
```

The problem here is that the returned value depends on the specific number entered as input; that is, it depends on the *value* of the x parameter. While this may seem like an extreme case, type-unstable functions often arise when processing user data or reading from some inputs.

Type stability of a function can be checked with the @code_warntype my_funct() macro. If you run @code_warntype f_unstable(2) in a terminal, you would obtain a lengthy output, but the important part is the first line, which would read as Body::Any (displayed in red in the Julia REPL). It is this information that tells you that the function is type-unstable.

One way to remove type instability in this example is to convert the elements of out_vector to be all of the same type:

```
function f_stable(x)    # Type stable
    out_vector = [1,convert(Int64,2.0),parse(Int64,"2")]
    if x < 0
        return out_vector[1]
    elseif x == 0
        return out_vector[2]
    else
        return out_vector[3]
    end
end
```

If you run @code_warntype f_stable(2), you receive in the first line of the output a Body::Float64 (displayed in a reassuring blue), which confirms the type stability of the function.

You can benchmark the two functions and see that the improvements due to type stability are huge:

```
@benchmark f_unstable(2) # median time: 234 ns
@benchmark f_stable(2)   # median time: 118 ns
```

Finally, if there are functions that really cannot be made type-stable, one strategy is to break the function into multiple definitions, where hopefully the computational-intensive part can be made type-stable.

### 8.1.4 Other Tips to Improve Performance

Here I present some solutions to problems that can severely limit Julia performance.

#### 8.1.4.1 Avoid Using Global Variables and Run Performance-Critical Code Within Functions

Global variables (variables that are defined outside any function) can change their type at any time. Instead of using a global variable inside a function, pass the variable as a function parameter. If this is not possible or practical, at least declare the global variable as const (to fix its type) or annotate its type at the point of usage in the function.

#### 8.1.4.2 Annotate the Type of Data Structures

As an exception to the rule that annotation in Julia is generally useless for performance, annotating the type of the inner elements of a data container with a concrete type can help with performance. Given that primitive types have fixed memory size, an array of them (e.g., Array{Float64}) can be stored in memory contiguously, in a highly efficient way. Conversely, containers of Any elements, or even any abstract type, are stored as pointers to the actual memory that ends up being sparsely stored and hence relatively slow to be accessed. So, rather than doing x = []; map(i->push!(x,i), 1:10000), do instead x = Int64[]; map(i->push!(x,i),1:10000) or, even better, preallocate the array: x = Vector{Int64}(undef,10000); map(i->x[i] = i, 1:10000).

## 8.1.4.3 Annotate the Fields of Composite Types

As a further exception to the annotation-is-not-needed rule, leaving the field of a structure unannotated (or annotated with an abstract type) will leave type-unstable all the functions that use that structure. If you want to keep the flexibility, use parametric structures instead. For example:

```
struct MyType2
  a::Float64
end
# ..or..
struct MyType3{T}
  a::T
end
# ..rather than..
struct MyType
  a
end
```

## 8.1.4.4 Loop Matrix Elements by Column and Then by Row

Matrix data is stored in memory as a concatenation of the various column arrays (*column-major* order), meaning the first column, followed by the second column, and so on. It is hence faster to loop matrices first over columns and then over their rows (so as to keep reading contiguous values):

```
M = rand(2000,2000)
function sum_row_col(M) # slower
    s = 0.0
```

```julia
    for r in 1:size(M)[1]
        for c in 1:size(M)[2]
            s += M[r,c]
        end
    end
    return s
end
function sum_col_row(M) # faster
    s = 0.0
    for c in 1:size(M)[2]
        for r in 1:size(M)[1]
            s += M[r,c]
        end
    end
    return s
end
@benchmark sum_row_col(M) # median time: 24.3 ms
@benchmark sum_col_row(M) # median time:  5.8 ms
```

## 8.2 Debugging

*Debugging* is, in general, "to remove bugs (= mistakes) from a computer program" (from the Cambridge dictionary).

Debugging is much harder in compiled languages, and this has prompted the development of many sophisticated, dedicated tools to help debugging programs.

The interactive and introspective nature of Julia makes debugging relatively simple, as the programmer can easily analyze, and eventually change, the code at any step. Still, specialized debugging tools have been

CHAPTER 8   EFFICIENTLY WRITE EFFICIENT CODE

developed to help debug code within functions that would otherwise be difficult to analyze in their individual components.

## 8.2.1 Introspection Tools

*Introspection* is the ability of a program to examine itself, and in particular, *type introspection* is the ability of a program to examine the type or properties of an object at runtime.

The first "introspection" tool is, in a wide sense, the Workspace panel, which is available in several Julia development environments (Figure 8-2 shows the one available in the VS Code IDE). The Workspace panel shows all the identifiers (variables, functions, and so on) available in the current scope and the objects bound to them.

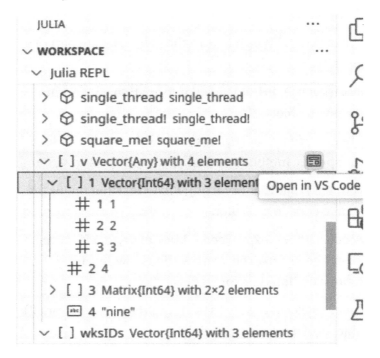

*Figure 8-2. Workspace panel in the VS Code IDE*

In addition, Julia offers the following introspection functions:

- `methods(myfunction)`: Retrieve all the function signatures; that is, all the different type of parameters the function accepts when it is called.
- `@which myfunction(myargs)`: Return which method would be used in a specific call (within the several available, as Julia supports multiple-dispatch, introduced in Chapter 3).
- `typeof(a)`: Return the type of a given object.
- `eltype(a)`: Return the type of the inner elements of a given object; for example, `typeof(ones(Int64,10))` is `Array{Int64,1}` and `eltype(ones(Int64,10))` is `Int64`.
- `fieldnames(AType)`: Return the fields that are part of a given type.
- `dump(myobj)`: Print detailed information concerning the given object.
- `@less myfunction(myargs)`: Show the source code of the specific method invoked.
- `@edit myfunction(myargs)`: Like `@less` but opens the source code in an editor.
- `@code_native expr`, `@code_llvm expr`, `@code_typed expr`, `@code_lowered expr`: Various low-level representations of expr.
- `names(MyModule,all=false)`: List the exported (or all) objects of module myModule.

- `sizeof(obj_or_type)`: Get the size in bytes of a certain object or type. For example, `sizeof(1.5)` is 4 because a `Float64` has 64 bits (i.e., 4 bytes).

- `bitstring(obj)`: Represent as a string the memory used by `obj`. For example, `bitstream(8)` returns a long string made of 60 zeros followed by "100".

## 8.2.2 Debugging Tools

Julia provides a full-featured debugger stack. The basic functionality is provided by the `JuliaInterpreter.jl` (https://github.com/JuliaDebug/JuliaInterpreter.jl) package, while the user interface is provided by the `Debugger.jl` (https://github.com/JuliaDebug/Debugger.jl) command-line package or the Julia VS Code extension itself.

Since these functionalities are similar, and since using a graphical user interface may be a bit more comfortable, I'll only discuss the integration of the debugger into VS Code. The VS Code integrated debugger allows developers to inspect the source code at any point and to set *breakpoints*, points where the execution stops so that the programmer can inspect the situation, change the code, or continue.

To start debugging Julia code, you can do either of the following:

- **Debug the whole file on a dedicated process**: Click the **Run** button and then click **Run and Debug** (see Figure 8-3).

- **Debug a specific Julia expression within a running session**: In the Julia terminal, type `@run expression_to_debug`. For example, `@run foo(5)`.

Both methods will execute the file or expression until a breakpoint (if any) is reached. Note that in order to debug a particular function call, you must have already evaluated the function.

CHAPTER 8   EFFICIENTLY WRITE EFFICIENT CODE

The code in Figure 8-3 illustrates the result of debugging the whole file without any breakpoints, which serves as a good example to debug.

***Figure 8-3.***  *Debugging session (start) in VS Code*

Rather than debugging the whole file, let's consider the more common situation in which you want to debug a single function call and you want to set some breakpoints. Figure 8-4 portrays a debugging session started by typing `@run foo(5)` in the Julia terminal, with the code halted on a breakpoint on the third line.

CHAPTER 8   EFFICIENTLY WRITE EFFICIENT CODE

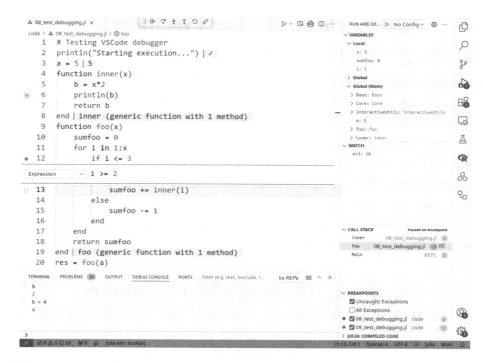

*Figure 8-4. Debugging session with breakpoint in VS Code*

## 8.2.3 Actions to Take Before Debugging (or While Execution Is Paused)

Before you start debugging (or while execution is paused), you can add or remove breakpoints by clicking the edge to the left of the code. Current breakpoints are marked with a red dot and listed in the Breakpoints panel, where you can edit them to add additional conditions, such as a hit count or an expression to evaluate at that particular point. In the Watch panel you can add arbitrary expressions to be evaluated, again at the point of the code that is currently paused.

## 8.2.4 Things to Inspect when Execution Is Paused

When execution is paused, you have access to the local variables and, by clicking Global (Main) in the Variables panel, to the global variables. To access the local variables in the parts of the code that lead to the current paused point, you can click the call stack. In the screenshot shown in Figure 8-4, I have clicked foo and thus gained access to the local variables i and sumfoo, which would otherwise be out of scope once inside the inner function. The Debug Console (at the bottom of the screenshot) is similar in many ways to the Watch panel, except that the expressions in the Watch panel are automatically reevaluated at each halt, while those in the Debug Console are reevaluated only once.

## 8.2.5 Debugging Choices when Execution Is Paused

The Debug Console has another important use: not only can you watch a local variable, you can also modify its object or rebind it to another variable. The screenshot in Figure 8-4 shows that I have rebounded b, which was previously bound to the value 2, to the value 4. When I continue to execute the code, b will remain bound to 4 until the function inner returns.

---

  **Caution**

To interact with the debugger, you must use the Debug Console, not the standard Julia terminal.

---

A floating toolbar above the code panel contains the graphical controls for the debugging session. From left to right, these controls (and their corresponding keypresses) are as follows:

- **Continue (F5)**: Continue execution until the debugged function terminates or a valid breakpoint is hit. `@run function(args)` is a shortcut for `@enter function(args)` plus pressing Continue.

- **Step Over (F10)**: Evaluate the current line and stop just before evaluating the next line.

- **Step Into (F11)**: Execute the next function call and stop at the beginning of the subfunction (e.g.. in Figure 8-4 it would call `println(b)` and stop at its beginning).

- **Step Out (Shift+F11)**: Continue execution until the current function exits or a valid breakpoint is hit.

- **Restart (Ctrl+Shift+F5)**: Restart debugging in the same way as it was started.

- **Stop (Shift+F5)**: Terminate the debugging session and return control to the normal REPL.

Finally, note that in a debugging session the code is interpreted rather than compiled, so the code will run much slower. However, you can set various options to tell the debugger to keep the parts of the code you are *not* debugging compiled. See the Julia for VS Code documentation for instructions.

## 8.3 Managing Runtime Errors (Exceptions)

In Julia, runtime errors can be handled, like in many other languages, with a `try/catch` block:

```
try
  # ..some dangerous code..
```

```
catch
    # ..what to do if an error happens, for example send an error
    message using:
    error("My detailed message")
end
```

You can also check for a *specific* type of exception. The following snippet uses exceptions to return `missing` if a specific key is not found in a dictionary representing some data:

```
data =Dict(("volume","Germany",2020) => 3683,
           ("volume","France",2020)  => 3055)
function volume(region, year)
    try
        return data["volume",region,year]
    catch e
        if isa(e, KeyError)
            return missing
        end
        rethrow(e)
    end
end
volume("Germany",2020) # 3683
volume("Germany",2025) # missing
```

Of course, in this example you could have just used the built-in function `get(data,("Volume","Germany",2025),missing)`.

# CHAPTER 9

# Parallel Computing in Julia

The following third-party packages are covered in this chapter:

| CUDA.jl | https://github.com/JuliaGPU/CUDA.jl | v5.4.3 |
| oneAPI.jl | https://github.com/JuliaGPU/oneAPI.jl | v1.6.0 |

This chapter follows up the discussion of efficiently writing efficient code from Chapter 8 by introducing a specific way of optimizing the execution of your code, namely *parallelization*.

Julia provides several packages for performing computations on the graphical processing unit (GPU), as well as core functionality to facilitate the parallelization of code on the CPU, either using multiple threads or using multiple processes. GPU parallelization is ideal when the same basic operations need to be performed independently on each element of a large data pool, and these operations can be chained together. Threads have the advantage of sharing the same memory (so there is no need to copy data across multiple threads) and using much more powerful hardware, but they are restricted to running on the same CPU. Processes each work on their own memory (so data must be copied between processes, which is a computational cost that must be balanced against the benefits of parallel execution), but they can be spawned on multiple CPUs or even multiple machines (e.g., through Secure Shell [SSH] connections).

# CHAPTER 9  PARALLEL COMPUTING IN JULIA

## 9.1 GPU Programming

GPU programming is available through the following vendor-specific packages, of which the NVIDIA GPU package is by far the most mature:

- NVIDIA: CUDA.jl (https://github.com/JuliaGPU/CUDA.jl)
- AMD: AMDGPU.jl (https://github.com/JuliaGPU/AMDGPU.jl)
- Intel: oneAPI.jl (https://github.com/JuliaGPU/oneAPI.jl)
- Apple: Metal.jl (https://github.com/JuliaGPU/Metal.jl)

These packages all provide different abstractions for programming on the GPU:

(a) high-level programming using specific abstract arrays that operate on the GPU (CuArray, ROCArray, oneArray, and MtlArray, respectively)

(b) functionality for writing custom GPU kernels

(c) low-level wrappers for the respective APIs

On top of these specific implementations, KernelAbstractions.jl (https://github.com/JuliaGPU/KernelAbstractions.jl) wraps the preceding packages to provide vendor-independent ways to define and invoke GPU kernels.

GPU programming is a broad topic, and a full treatment is beyond the scope of this book, so in this section I present an example of using the high-level interface of CUDA.jl to compute the forward connectivity of a custom deep neural network with three layers. Neural networks are powerful machine learning algorithms that, as you will see, benefit enormously from GPU programming. Neural networks are discussed in detail in Chapter 12, but this example will make you appreciate their simplicity, which is astonishing compared to their power.

The high-level interface of CUDA consists of simply replacing the standard Array with CuArray. Provided you can implement your algorithm in vectorized form (including broadcasting) and with functions that work on preallocated arrays (i.e., the functions return nothing), you have nothing else to do. In particular, you don't need to define GPU kernels manually.

The following code snippet defines the neural network functions: the so-called activation function, which is applied scalarly to the layer outputs; the individual forward_layer() function, which is a linear combination of the input values with the layer weights; and the overall forward_network!() function, which is simply a composition of the individual layer functions and works on a preallocated output.

```
# Function definitions
relu(x) = max(0,x)                          # Activation function
forward_layer(x,w,w0,f) = f.(w*x .+ w0)     # Individual layer
function forward_network!(y,x,w1,w2,w3,w01,w02,w03,f=relu)
# Whole chain
    x1 = forward_layer(x,w1,w01,f)
    x2 = forward_layer(x1,w2,w02,f)
    y .= forward_layer(x2,w3,w03,identity)
    return nothing
end
```

You can now create some random data and random weights to test the network on the CPU:

```
# CPU data
(nd0,nd1,nd2,ndy) = (20000,30000,30000,1)
x   = rand(Float32,nd0);      y = Vector{Float32}(undef,ndy)
```

```
w1  = rand(Float32,nd1,nd0); w2 = rand(Float32,nd2,nd1);
w3  = rand(Float32,ndy,nd2)
w01 = rand(Float32,nd1);     w02 = rand(Float32,nd2);
w03 = rand(Float32,ndy);
# CPU call
forward_network!(y,x,w1,w2,w3,w01,w02,w03,relu)
```

So far you haven't done anything special, and the network was running on the CPU. To run on the GPU, you simply preallocate the output GPU array with CuArray{Float32}(undef,ndy) and create a "GPU" version of the input data and layer weights:

```
# GPU data
y_g  = CuArray{Float32}(undef,ndy)
x_g  = CuArray(x)
w1_g = CuArray(w1);   w2_g = CuArray(w2);
w3_g = CuArray(w3);
w01_g = CuArray(w01); w02_g = CuArray(w02);
w03_g = CuArray(w03);
```

At this point you can simply call the function with this CuArray. Note that you don't need to change anything in the function definitions themselves; they are automatically compiled and scheduled as GPU kernels.

```
# GPU call
forward_network!(y_g,x_g,w1_g,w2_g,w3_g,w01_g,w02_g,w03_g,relu)
```

Finally, you can compare the correctness of the GPU computation with the CPU computation as a reference. To return a normal array from a GPU array, use Array(gpu_array):

```
# Correctness check
@test y ≈ Array(y_g)
```

While the code in the preceding example is specific to the NVIDIA GPU, adapting it to the other GPU vendors is trivial. Just replace `CuArray()` with `ROCArray()`, `oneArray()`, or `MtlArray()`.

## 9.1.1 Benchmarking CPU vs. GPU

If you followed the steps in the previous sentence, you now are running the code on the GPU. But what is the performance gain? To give you an idea, I have benchmarked the `forward_network!` function in different situations, ranging from a powerful server with a modern NVIDIA GPU and to an old Ultrabook with a built-in graphics card (using the `oneAPI` package); on both `Float64` and `Float32` data; and on small and large inputs. I got the following benchmark table:

| | Server CPU: AMD Threadripper PRO 5975WX | Server GPU: NVIDIA RTX A4500 | Laptop CPU: Intel i5-8350U | Laptop GPU: UHD Graphics 620 |
|---|---|---|---|---|
| Small (nd0,nd1,nd2 = 200,300,300) | | | | |
| 32 bits | 352 μs | 181 μs | 122 μs | 2866 μs |
| 64 bits | 345 μs | 212 μs | 189 μs | 3372 μs |
| Large (nd0,nd1,nd2 = 20000,30000,30000) | | | | |
| 32 bits | 107 ms | 0.20 ms | 434 ms | 278 ms |

There are three notable points about such benchmarks. The first is that the difference between `Float32` and `Float64` in terms of execution speed is minimal, if any, for the GPUs that support native `Float64` computation. However, an important advantage of using `Float32` is the lower memory consumption. GPU memory is still much smaller than CPU memory and is often the bottleneck. On my systems, I ran out of GPU memory when trying to run the benchmark for the large system on `Float64`.

The second aspect to note is how fast GPU computing has evolved in comparison to CPU computing. While modern CPUs are undoubtedly faster than older systems for large datasets, for GPUs the difference can be several orders of magnitude faster.

The third, and perhaps most important, conclusion is that GPU computing is most valuable for large array sizes. There are indeed some "fixed costs" of GPU computation, but these costs decrease very little with array size, as you can see in the previous benchmark table. This is all the more true because I didn't include the very important times for converting the arrays to `CuArray` in the benchmark; I included only the execution of the `forward_network!` function, which works on preallocated data.

One strategy might then be to try to keep the computations on the GPU without converting to and from the CPU. In the neural network example, this would mean that the forward, backward, gradient, and update passages would all live on the GPU. Fortunately, once the computation can be expressed in vector form, the high-level GPU API provided by the Julia GPU packages helps to achieve this goal.

## 9.2 Multithreading (on the CPU)

Another way of parallelizing computation is to use threads. They don't require any additional hardware, and they eliminate the cost of copying memory. Threads functionality has been available in the `Threads` standard library since Julia v1.3.

Julia needs to know the number of threads to use at startup, so it is not possible to add threads to a running Julia process. You can either start Julia with the parameter -t desired_threads on the command line or use the VS Code Julia extension setting Julia: Num Threads. While you can set any positive integer, you would normally consider the number of available CPU cores as the upper limit of this setting (equivalent to -t auto). Once inside Julia, you can check the number of threads available to Julia with Threads.nthreads() and get the thread ID of the "current" thread with Threads.threadid().

As with GPU programming, Julia provides not only all the tools necessary for multithreaded programming in a flexible and efficient way, but also a high-level macro, Threads.@threads. You can use this macro to parallelize for loops where each iteration is independent of the other iterations (e.g., it works on a different element of the container being iterated), and the order of execution is not important. It is sufficient to place this macro before the for loop to run it in parallel.

In the following example, you first define a function to run element-wise, and then you define the outer function with the for loop, which can be multithreaded or not. This allows you to compare the different speedups that multithreading can give you if the element-wise operations to be parallelized have different computational costs. Finally, check that the multithreaded example gives you the same result as the scalar implementation.

```
function inner_function(x,cheap) # sum of square roots
    N = cheap ? 10 : 10000
    return sum(sqrt(x) for i in 1:N)
end
function rootsq_sums!(x;multithread=false,cheap=true)
    if multithread
        Threads.@threads for i in eachindex(x)
            x[i] = inner_function(x[i],cheap)
        end
```

## CHAPTER 9    PARALLEL COMPUTING IN JULIA

```
    else
        for i in eachindex(x)
            x[i] = inner_function(x[i],cheap)
        end
    end
    return x
end

@assert rootsq_sums!(collect(1.0:100.0)) == rootsq_sums!
(collect(1.0:100.0),multithread=true)
```

I have benchmarked `rootsq_sums!`, the results of which are presented in the following table:

|  | Single Threads | Multiple Threads (8) | Speedup |
|---|---|---|---|
| Small data (x=1.0:100.0) | | | |
| Cheap element-wise op | 0.36 µs | 5.18 µs | 0.07x |
| Expensive element-wise op | 1379 µs | 147 µs | 9.3x |
| Large data (x=1.0:10000.0) | | | |
| Cheap element-wise op | 0.03 ms | 0.01 ms | 3.0x |
| Expensive element-wise op | 113 ms | 14 ms | 8.1x |

As you can see, even with multithreading there are some fixed costs of parallelization, but compared to the GPU, these are lower. More importantly, the gain can be large with relatively few elements to parallelize, and it remains large even if the operation being parallelized is computationally expensive.[1]

---
[1] To be fair, the multithreading macro may have enabled other low-level optimizations that the LLVM compiler didn't apply by default in this case.

## 9.3 Multiprocessing

All the functions presented in this section are part of the `Distributed` standard library package.

### 9.3.1 Adding and Removing Processes

Unlike threads, processes can be added and removed at any time:

```
wksIDs = addprocs(3) # 2,3,4

println("Worker pids: ")
for pid in workers()
    println(pid) # 2,3,4
end

rmprocs(wksIDs[2]) # or rmprocs(workers()[2]) remove process pid 3

println("Worker pids: ")
for pid in workers()
    println(pid) # 2,4 left
end

@everywhere println(myid()) # 2,4
```

The first row adds three processes. In this case, it adds them on the same machine as the one running the main Julia process, but it is possible to specify in `addprocs()` the SSH connection details directly in order to add the processes to other machines (Julia must be installed on those machines as well). Generally, it makes sense for *n* to equal the number of cores available on the machine. It is also possible to start Julia directly with multiple processes using the command line flag -p (e.g., ./julia -p 2). The -p argument implicitly loads the `Distributed` module.

In this example, the "internal ids" of the new processes (different from the OS-level assigned pids) are saved in the array `wksIDs`, although you don't really need it, as a call to `workers()` would return that array as well. The main Julia process (e.g., the one providing the interactive prompt) always has a pid of 1 and, unless it is the sole process, it is not considered a worker.

You can then check the working processes that are active and remove those that you no longer need with `rmprocs(pid)`.

Finally, you can use the function `myid()` to return the pid of a process and spread the function to print it across all the working processes using the `@everywhere` macro.

## 9.3.2 Running Heavy Computations on a List of Items

This section and the next section present two common patterns of parallelization. The pattern presented in this section involves running a computationally expensive operation over a list of items, where you are interested in the result of each of them. Critically, the computation of each item is independent of the other items.

As an example, you are going to reuse the `fib` function you saw earlier. You first need to be sure that all processes (assume you added three working processes with `addprocs(3)`) know about the function you want them to use. You can use the `@everywhere` macro in front of the definition of the function in order to transfer the function to all the workers (but if you add further processes at a later stage, they will not have access to such a function).

If you have multiple functions to invoke in the working threads (or there are several subfunctions to call), a convenient option is to define all of them and the eventual global variables they need in a `begin` block prefixed with the `@everywhere` macro (i.e., `@everywhere begin [shared functions definitions] end`) or put them in a file and run `@everywhere include("computationalcode.jl")`.

You can then create the array of input data and run the heavy operation on each element of it with `pmap(op,inputData)`:

```
a = rand(1:35,100)
@everywhere function fib(n)
    if n == 0 return 0 end
    if n == 1 return 1 end
    return fib(n-1) + fib(n-2)
end
@benchmark results = map(fib,a)   # serialized: median
                                    time:    490.473 ms
@benchmark results = pmap(fib,a)  # parallelized: median time:
                                    249.295 ms
```

The `pmap` function automatically and dynamically picks up the "free" processes, assigns them the job, and merges the results in the returned array. Only the evaluation of the function is done in the workers. `pmap` is convenient when the computation cost of the function is high, like in this case. Note, however, that even if you are using three working threads, computational time is divided by a factor well below 3[2].

## 9.3.3 Aggregate Results

The second common pattern of parallelization is when you want to perform a small operation on each of the items but you also want to perform an "aggregation function" at the end to retrieve a scalar value (or an array if the input is a matrix).

---

[2] This is known in computer science as *Amdahl's Law*, and it is due to bottlenecks from the serialized part of the code and the computational costs of the parallelization.

In these cases you can use the @distributed (aggregationFunction) for construct. As an example, you run in parallel a division by 2 and then use the sum as the aggregation function (assume three working processes are available):

```
function f(n)
  s = 0.0
  for i = 1:n
    s += i/2
  end
    return s
end
function pf(n)
  s = @distributed (+) for i = 1:n # aggregate using sum on
                                   variable s
      i/2                          # last element of for cycle
                                   is used by the aggregator
  end
    return s
end
@benchmark  f(10000000)  # median time:      11.478 ms
@benchmark pf(10000000)  # median time:       4.458 ms
```

The two patterns presented are some of the most frequently encountered in code parallelization. Julia offers a fairly complete list of high- and low-level functions for parallelization, and some external packages, like DistributedArrays.jl (https://github.com/JuliaParallel/DistributedArrays.jl), help you to achieve efficient parallelized code for many situations in a relatively simple way.

# PART II

# Packages Ecosystem

# PART II

# Packages Ecosystem

# CHAPTER 10

# Working with Data

The following third-party packages are covered in this chapter:

| | | |
|---|---|---|
| DataFrames.jl | https://github.com/JuliaData/DataFrames.jl | v1.6.1 |
| DataFramesMeta.jl | https://github.com/JuliaStats/DataFramesMeta.jl | v0.15.3 |
| Query.jl | https://github.com/queryverse/Query.jl | v1.0.0 |
| JLD2.jl | https://github.com/JuliaIO/JLD2.jl | v0.4.51 |
| HDF5.jl | https://github.com/JuliaIO/HDF5.jl | v0.17.2 |
| IndexedTables.jl | https://github.com/JuliaComputing/IndexedTables.jl | v1.0.1 |
| CategoricalArrays.jl | https://github.com/JuliaData/CategoricalArrays.jl | v0.10.8 |
| Pipe.jl | https://github.com/oxinabox/Pipe.jl | v1.3.0 |
| Plots.jl | https://github.com/JuliaPlots/Plots.jl | v1.40.5 |
| StatsPlots.jl | https://github.com/JuliaPlots/StatsPlots.jl | v0.15.7 |

# CHAPTER 10   WORKING WITH DATA

Although Julia already natively provides a fast structure for working with tabular (and multidimensional) data, the parameterized type Array{T,n}, a very large collection of third-party packages, has been created to work specifically with numerical data. This was created, broadly speaking, because of two needs:

- **Greater convenience**: Packages such as DataFrames.jl and IndexedTables.jl add metadata like column names, allowing developers to retrieve data by name instead of by position, query it with Query.jl or DataFramesMeta.ij, and then join, group, and chain operations with Pipe.jl, and so on.

- **Efficient handling of heterogeneous data**: In a Julia Base Array{T,2}, all data in the matrix must be of the same type for efficiency. The alternative is to work with Array{Any,2} and accept lower performance.

With the aforementioned packages, different columns can be of different types, a situation quite common when working with datasets, and the computations involving these data structures remain performant.

This chapter introduces the main third-party Julia packages that include tools for manipulating data (including data structures), querying data, plotting data, and visualizing data.

---

 **Note**

DataFrame column names are stored internally as Julia symbols, not strings. The DataFrames API has evolved to use strings, but in many functions you can still use symbols instead of strings to refer to column names if you prefer.

---

## 10.1 Using the DataFrames.jl Package

DataFrames are the first data structure you will see, and perhaps the most used. They are very similar to R's data frames and Python's pandas. The approach and function names are also very similar, although the way the API is accessed may be slightly different. Unlike pandas, the `DataFrames.jl` package only works with two-dimensional (tabular) data.

Internally, a `DataFrame` is a collection of standard arrays, each of its own type T, or possibly of type `Union{T,Missing}` if some data is unavailable.

If you are coming from R and prefer the tidyverse approach, have a look at `Tidier.jl` (https://github.com/TidierOrg/Tidier.jl), which wraps `DataFrames` and other data analysis tools.

### 10.1.1 Installing and Importing the Library

You can install and load the library in the usual way:

- Install the library: `] add DataFrames`
- Load the library: `using DataFrames`

### 10.1.2 Creating a DataFrame or Loading Data

There are different ways to create a DataFrame ("df" for short) and loading data. Methods for loading a df from a comma-separated, Excel, or ODS file were discussed in Chapter 5.

You can also create a df from scratch:

**Note**

The following (fictional) database of timber markets will be referenced throughout this chapter.

---

```
df = DataFrame(region      = ["US","US","US","US","EU","EU",
                              "EU","EU"],
               product     = ["Hardwood","Hardwood","Softwood",
                              "Softwood","Hardwood","Hardwood",
                              "Softwood","Softwood"],
               year        = [2010,2020,2010,2020,2010,202
                              0,2010,2020],
               production  = [3.3,3.2,2.3,2.1,2.7,2.8,1.5,1.3],
               consumption = [4.3,7.4,2.5,9.8,3.2,4.3,6.5,3.0])
```
---

To create an empty df, you can use

```
df2 = DataFrame(A = Int64[], B = Float64[])
```

To convert a df from a matrix of data, you must either pass a second parameter as a header name or use the :auto keyword:

---
```
mat = [1 2 3; 4 5 6]
headers = ["c1", "c2", "c3"]
df2 = DataFrame(mat,:auto) # by default col names are x1,
                            x2, x3...
df2 = DataFrame(mat,headers)
```
---

It is also possible to create a df from a table that is hard-coded:

CHAPTER 10 WORKING WITH DATA

```
using CSV
df2 = CSV.read(IOBuffer("""
region product  year production consumption
US      Hardwood 2010 3.3       4.3
US      Hardwood 2020 3.2       7.4
US      Softwood 2010 2.3       2.5
US      Softwood 2020 2.1       9.8
EU      Hardwood 2010 2.7       3.2
EU      Hardwood 2020 2.8       4.3
EU      Softwood 2010 1.5       6.5
EU      Softwood 2020 1.3       3.0
"""), DataFrame, delim=" ", ignorerepeated=true)
```

## 10.1.3 Gaining Insight into the Data

Once you have a DataFrame, you may want to analyze its data or structure by using the following functions:

- `show(df)`: Show an extract of the df (depending on the space available). Use `allrows=true` and `allcols=true` to get the whole df, whatever its size.
- `first(df,n)`: Display the first n lines of the df.
- `last(df,n)`: Display the last n lines of the df.
- `describe(df)`: Return the type and basic descriptive statistics for each column of the df.
- `names(df)`: Return an array of column names (as strings). You can optionally filter the eltype; for example, filter with `names(df,AbstractString)` to return only the names of columns whose elements are AbstractString or a subtype of it.

189

- `for r in eachrow(df)`: Iterate over each row.
- `for c in eachcol(df)`: Iterate over each column.
- `unique(df.fieldName)` or `[unique(c) for c in eachcol(df)]`: Return the unique elements of the column(s).
- `[eltype(c) for c in eachcol(df)]`: Return an array of column types.
- `size(df)`: Return the size of the df in terms of (number of rows, number of cols).

## 10.1.4 Filtering Data (Selecting or Querying Data)

To select (filter) data, you can use the methods provided by the `DataFrames.jl` package or specialized methods provided by other packages, such as `DataFramesMeta.jl` or `Query.jl`.

In particular, the `Query.jl` package aims to provide a generic API that is valid regardless of the data back end.

- To select whole columns:
  - `df[:,cNames]`: Select multiple columns (e.g., `df[:,["product","year"] ])` by copying the data.
  - `df[!,cNames]`: Select multiple columns (e.g., `df[!,["product","year"] ])` by referencing the data.
  - `df.cName`: Select a single column (e.g., `df.product`) (referenced, i.e. equal to `df[!,cName]`).
  - `df[:,cPos]` and `df[!,cPos]`: Select one or more columns (e.g., `df[:,2]`) by respectively copying or referenceing the data.

- To select whole rows:
  - df[rPos,:]: Select one or more rows (e.g., df[2,:]). Unlike columns, rows do not have a label.
- To select cells:
  - df.cName[rPos]: Select specific cells in the cName column (e.g., df.product[1:2:6]).
  - df[rPos, cName(s)] or df[rPos,cPos]: Select specific cells by row position and column name or position (e.g., df[2,"product"]).

Where:

- cName is a single column name (string but symbol is also accepted).
- cNames is an array of column names (strings or symbols).
- rPos are the row positions (scalar, array, or range).
- cPos are the column positions (scalar, array, or range).

The selection returned is

- An Array{T,1} if it is a single column
- A DataFrameRow (similar to a DataFrame) if it is a single row
- T if it is a single cell
- Another DataFrame otherwise

Instead of selecting rows by position, you can select them by *boolean selection* (introduced in Chapter 2)—that is, by an array of congruent size, where the individual true values correspond to the selected rows.

This also works for column positions, to select rows and columns at the same time (but not for a matrix of boolean values to use as a mask). The following are some examples:

- `df[ [true, true, false, false, false, false, false, false], [true, true, true, false, true]]` selects the first two rows of all columns except the fourth.

- `df[ df.year .>= 2020, :]` filters the rows by value, based on whether the value meets a certain condition (here the year is at least 2020). Note that `df.year` is a vector, so you need to broadcast the condition. Checking the condition will result in a vector of boolean values that will be used for the selection.

- `df[ (df.year .>= 2020) .& in.(df.region, Ref(["US","China"])), :]` filters by multiple conditions in the boolean selection. The dot is needed to vectorize the operation. Note the use of the bitwise "and" (the single ampersand) and the use of the `Ref` function, which returns a 0-dimension container that protects its contents from being broadcast.

- `df[startswith.(df.region,"E"),:]` filters based on the initial pattern of desired values.

Alternatively, filter rows can be obtained with `@subset(df, condition(s))` from the `DataFrameMeta.jl` package, where in the case of multiple conditions, all must be checked. If the column name is stored in a variable, you must wrap it using the `cols()` function. For example:

`reg_tofilter = :region; @subset(df, :production .> 3, cols(reg_tofilter) .== "US").`

Note that you have to use symbols as column names in `@subset`.

A third (and perhaps more elegant, although longer) way to filter a dataframe is to use the Query.jl (https://github.com/queryverse/Query.jl) package. The first example selects a subset of both rows and columns; the second shows how you can mix multiple selection criteria:

```
df2 = @from i in df begin
         @where i.region == "US"
         # Select a group of columns, eventually changing
         their name:
         @select {i.product, i.year, USProduction=i.
         production}
         @collect DataFrame
      end
df2 = @from i in df begin
         @where i.production >= 3 && i.region in
         ["US","China"]
         @select i # Select the whole rows
         @collect DataFrame
      end
```

## 10.1.5 Editing Data

There are many ways to edit data:

- `df[rowIdx,col1] .= aValue`: Edit values by specifying rows and columns (e.g., `df[[1,2],"production"] .= 4.2`).

- df[(df.col1 .== "foo"), col2] .= aValue:
  Change values by filtering by a boolean selection (e.g.,
  df[(df.region .== "US") .& (df.product .==
  "Hardwood"), "production"] .= 5.2).

- df.col1 = map(akey->my_dict[akey], df.col1):
  Replace values (or add columns) based on a dictionary
  (e.g., reg_fullnames = Dict("US" => "United
  States", "EU" => "European Union"); df.region
  = map(r_shortname->reg_fullnames[r_shortname],
  df.region)).

- df.a = df.b .* df.c: Calculate the value of an
  existing or new column based on the values of other
  columns (e.g., df.net_export = df.production .-
  df.consumption). Remember that you are working
  with vectors, so you need to broadcast the operation.
  Alternatively, you can use map; for example, df.net_
  export = map((p,c) -> p - c, df.production,
  df.consumption).

- push!(df, [a_row_vector]): Append a row to the df
  (e.g., push!(df,["EU" "Softwood" 2012 5.2 6.2])).

- deleterows!(df,rowIdx): Delete the given
  rows. rowIdx can be a scalar, a vector, or a
  range. Alternatively, you can just copy the df
  without the unneeded rows (e.g., i=3; df2 =
  df[[1:(i-1);(i+1):end],:]).

- df = similar(df,0): Empty a df. df2= similar(df,n)
  copies the structure of a df and adds n rows (with
  garbage data) to the new DataFrame (n defaults to
  size(df)[1]). You can use it with n=0 to empty a
  given df.

### 10.1.5.1 Sorting

Dataframes can be sorted by multiple columns with `sort!(df, columnsToSort, rev = [reverseSortingFlags])`; for example, `sort!(df, ["year","product"], rev=[false,true])`. The (optional) reverse order parameter (`rev`) must be an array of boolean values of the same size as the `cols` parameter. Custom sorting of categorical variables can be obtained using the `CategoricalArrays.jl` package described later.

## 10.1.6 Editing the Structure

There are also many ways to edit the structure of the data:

- `select!(df,Not([col1,col2]))`: Delete columns by name.
- `rename!(df, [newCName1,newCName2,newCName3])`: Rename (all) columns. Use `rename!(df, Dict(oldCName1 => newCName1))` to rename only a few columns.
- `df = df[:,[b, a] ]`: Change the order of the columns.
- `df.id = 1:size(df, 1)`: Add an `id` column (useful for unstacking, an operation you'll see later in this chapter).
- `df.a = Array{Union{Missing,Float64},1}(missing,size(df,1))`: Add a `Float64` column (filled with missing values by default).
- `insertcols!(df, i, colName1=>colContent1, colName2=>colContent2)`: Insert columns at position `i` (or at the end if `i` is omitted).

- `df.a = convert(Array{T,1},df.a`: Convert column type to T.
- `df.a = map(string, df.a)`: Convert from `Int64` (or `Float64`) to `String`.
- `string_to_float(str) = try parse(Float64, str) catch; return(missing) end; df.a = map(string_to_float, df.a)`: Convert from `String` to `Float64` (converting to integers is similar).
- `df2 = similar(df1, n)`: Copy the structure of a DataFrame (to an empty one if n is 0; otherwise, with n rows of undefined data).

## 10.1.6.1 Merging/Joining/Copying Datasets

You can use the following methods to merge or join datasets. If the resulting dataframe may end up with duplicate column names, use `makeunique=true` to avoid an error.

- `df = vcat(df1,df2,df3)`: Vertically concatenate different dataframes with the same structure (or `df = vcat([df1,df2,df3]...)` (note the three-dot splat operator at the end).
- `df = hcat(df1,df2)`: Horizontally concatenate dataframes with the same number of rows.
- `df = innerjoin(df1, df2, on = commonCol)`: Join dataframes horizontally. It merges horizontally and returns only the rows with matching keys on both sides.
- `df = leftjoin(df1, df2, on = commonCol)`: Join dataframes horizontally. It returns all rows from the left df, possibly extended with matching rows from the right df (otherwise `missing` is filled).

- `df = rightjoin(df1, df2, on = commonCol)`: Join dataframes horizontally. It returns all rows from the right df, possibly extended with matching rows from the left df (otherwise `missing` is filled).
- `df = outerjoin(df1, df2, on = commonCol)`: Join dataframes horizontally. It returns all rows from both dataframes. If a common key is found, the row is merged. Otherwise, the columns corresponding to the other df are filled with `missing`.
- `df = semijoin(df1, df2, on = commonCol)`: Join dataframes horizontally. It returns only the rows with matching keys (like `innerjoin`), but only the fields from the left df.
- `df = antijoin(df1, df2, on = commonCol)`: Join dataframes horizontally. It returns only the rows from the left df that don't have a match in the right df.
- `df = crossjoin(df1, df2)`: Join dataframes horizontally. It joins every single row of the left df with every single row of the right df.

The `on` parameter can also be an array of common columns to use as matching keys. Note that if more than one matching row is found when joining a single row, all matching rows are returned for that row.

## 10.1.7 Working with Categorical Data

Often a dataset contains fields with a limited number of possible values, and there is no natural ranking between the different values. Examples include fields for selecting the gender of a group of people or the country of origin of a product. These types of variables are categorical, and `CategoricalArrays.jl` (https://github.com/JuliaData/CategoricalArrays.jl) is a package for representing them efficiently.

## CHAPTER 10  WORKING WITH DATA

You can create a categorical array from a dense representation with `categorical(an_array)`; for example, `df.region = categorical(df.region)`. Note that although filtering with categorical values reduces the amount of memory required to store the data, it is not necessarily faster (in fact, it can be slightly slower).

You can extract the (unique) levels of the categorical array with `levels(an_array)` and give them an arbitrary order (e.g., for printing) with `levels!(an_array,[level_desired_1,level_desired_2,...])`. To rename the levels, simply modify the vector returned by `levels`. Here is an example:

```
using CategoricalArrays
df.region = categorical(df.region) # Transformation to a
categorical array
levels(df.region)  # ["EU", "US"]
sort!(df,"region") # EU rows first
levels!(df.region,["US","EU"]) # Providing a personalized order
levels(df.region)  # ["US", "EU"]
levels(df.region) .= ["United States", "European Union"]
# Renaming
sort!(df,"region") # United States rows first
```

> **ⓘ Note**
>
> While I describe the `CategoricalArrays` package here in the context of dataframes, `CategoricalArrays` works with any `Array`, including matrices.

To convert a categorical array back into a regular one, use `unwrap` element-wise; for example, `df.region = unwrap(df.region)`.

## 10.1.8 Managing Missing Values

Recall that missing values (i.e., `missing`, the only instance of the singleton type `Missing`) will propagate silently through operations if they aren't dealt with. Julia Base and the `DataFrames` package provide a number of methods for programmers to work with missing data and decide what to do with it, depending on the interpretation of "missingness":

- `allowmissing(df,colName)` and `allowmissing(an_array)`: Allow an array or a specific df column (or the whole df if colName is omitted) to store missing data along with its original data type. For dataframes, you can also use the in-place version `allowmissing!`.

- `disallowmissing(df,colName)` and `disallowmissing(an_array)`: The opposite of `allowmissing`, they prevent an array or df from holding missing values.

- `dropmissing(df)` and `dropmissing!(df)`: Return a df with only complete rows—that is, without `missing` in any field (`dropmissing(df,[col1,col2])` is also available). Within an operation (e.g., a sum), you can use `dropmissing()` to skip missing values before the operation. Note that `dropmissing` automatically performs a `disallowmissing`.

- `completecases(df)` or `completecases(df,["col1","col2"])`: Return a boolean array of the rows or specified fields that don't have a missing value.

- `nonmissingtype(Union{T,Missing})`: Return T, the type in the union that is not `Missing` (e.g., `nonmissingtype(Union{Float64,Missing})` returns `Float64`).

## CHAPTER 10 WORKING WITH DATA

- `isequal(a,b)`: Compare without missing propagation (e.g., `isequal("US",missing)` returns false, while `"US" == missing` would be neither `true` nor `false`, but `missing`).
- `ismissing(a)`: Check if `a` is `missing`.
- `df[ismissing.(df[!,i]), i] .= 0 for i in names(df, Union{Missing,Number})]`: Replace `missing` with 0 values in all numeric columns, such as `Float64` and `Int64`.
- `[df[ismissing.(df[!,i]), i] .= "" for i in names(df, Union{Missing,AbstractString})]`: Replace `missing` with `""` values (empty string) in all string columns.
- `n_missings = length(findall(x -> ismissing(x), array))`: Count the number of missing values in an array.

If you want to impute missing data, the `Imputation` module of `BetaML.jl` provides four different imputation models:

- `SimpleImputer`: Imputes data using the mean of the features (columns), optionally normalized by the p-norm of the records (rows). This model is the fastest.
- `GaussianMixtureImputer`: Imputes data using a generative (Gaussian) mixture model. This model offers a good compromise.
- `RandomForestImputer`: Imputes missing data using random forests, with optional replicable multiple imputation. This model is the most accurate.

CHAPTER 10 WORKING WITH DATA

- GeneralImputer: Imputes missing data using a vector (one per column) of arbitrary learning models (classifiers/regressors) implementing m = Model([options]), fit!(m,X,Y) and predict(m,X) (not necessarily from BetaML).

All these imputers work on arrays rather than dataframes, so you will need to convert your data to use them. For example:

```
using CSV, DataFrames,BetaML
# Build a dataframe with some missing data
df2 = CSV.read(IOBuffer("""
   col1     col2      col3
   1.4      2.5       "a"
missing    20.5       "b"
   0.6       18    missing
   0.7      22.8      "b"
   0.4   missing      "b"
   1.6      3.7       "a"
"""), DataFrame, delim=" ", missingstring="missing",ignorerepeated=true)
# Impute the missing values
data_imputed = fit!(RandomForestImputer(),Matrix(df2))
# Copy back the imputed value into the original dataframe
[df2[:,c] .= data_imputed[:,c] for c in axes(df2,2)]
```

BetaML.jl and its API are discussed in detail in Chapter 12.

201

## 10.1.9 Pivoting Data

Table data can be represented equivalently using two types of layout: long and wide. In a *long* layout, you specify each dimension in a separate column, and then you have a single column for what you consider to be the "value." If you have multiple variables in the database, the "variable" also becomes a dimension column.

For example, using the timber market example, Table 10-1 shows a long-formatted table.

*Table 10-1.* *A Long-Formatted Table*

| Region | Product | Year | Variable | Value |
|---|---|---|---|---|
| US | Hardwood | 2010 | production | 3.3 |
| US | Hardwood | 2010 | production | 3.3 |
| US | Hardwood | 2020 | production | 3.2 |
| US | Softwood | 2010 | production | 2.3 |
| ..etc.. | | | | |

The long format is less human-readable, but it is very easy to work with and analyze.

The *wide* format, on the other hand, expresses values in multiple columns, with one or more dimensions expressed along the horizontal axis. See Table 10-2.

*Table 10-2. A Widely Formatted Table*

| Region | Product  | Year | Production | Consumption |
|--------|----------|------|------------|-------------|
| US     | Hardwood | 2010 | 3.3        | 4.3         |
| US     | Hardwood | 2020 | 3.2        | 7.4         |
| US     | Softwood | 2010 | 2.3        | 2.5         |
| US     | Softwood | 2020 | 2.1        | 9.8         |
| ..etc.. |         |      |            |             |

The timber example in the original format is a wide table, but you can create even wider tables using multiple horizontal axes (not supported by the `DataFrames` package). See Table 10-3.

*Table 10-3. An Even More Widely Formatted Table (with Multiple Horizontal Axes)*

|    |      | Production | | Consumption | |
|----|------|----------|----------|----------|----------|
|    |      | Hardwood | Softwood | Hardwood | Softwood |
| US | 2010 | 3.3      | 2.3      | 4.3      | 2.5      |
| US | 2020 | 3.2      | 2.1      | 7.4      | 9.8      |
| EU | 2010 | 2.7      | 1.5      | 3.2      | 6.5      |
| EU | 2020 | 2.8      | 1.3      | 4.3      | 3        |

The operation of moving axes between the horizontal and the vertical dimensions is called *pivoting*. In particular, *stacking* moves columns and expands them as new rows, thus moving from a wide format to a long format. *Unstacking* does the opposite, taking rows and placing their values as new columns, thus creating a "wider" table.

## 10.1.9.1 Stacking Columns

The `DataFrames` package provides the `stack(df,[cols])` function, in which you specify the columns that you want to stack. Use it without column names to automatically stack all floating-point columns. Note that the headers of the stacked columns are inserted as data in a `variable` column and the corresponding values in a `value` column.

---
```
long_df = stack(df,["production","consumption"])
long_df1 = stack(df)
long_df == long_df1    # true
 | Row | variable   | value   | region | product  | year  |
 |     | Symbol     | Float64 | String | String   | Int64 |
 |-----|------------|---------|--------|----------|-------|
 | 1   | production | 3.3     | US     | Hardwood | 2010  |
 | 2   | production | 3.2     | US     | Hardwood | 2020  |
 | 3   | production | 2.3     | US     | Softwood | 2010  |
 | 4   | production | 2.1     | US     | Softwood | 2020  |
 ...etc...
```
---

## 10.1.9.2 Unstacking

The `DataFrames` package also provides `unstack(longDf, [rowField(s)], variableCol, valueCol)`, in which you specify the list of columns that should remain as categorical variables even in the wide format, the name of the column that stores the category to be expanded on the horizontal axis, and the name of the column that contains the corresponding values.

You can omit the list of categorical variables with `unstack(longDf, variableCol, valueCol)`, in which case all existing columns except the one defining column names and the one defining column values will be retained as categorical columns:

CHAPTER 10 WORKING WITH DATA

```
wide_df = unstack(long_df,["region", "product",
"year"],"variable","value")
wide_df1 = unstack(long_df,"variable","value")
wide_df == wide_df1 # true
```

## 🛈 Note

**Multiple Axis Tables**

The fact that the `variable` parameter of `unstack` is a scalar and not a vector is the result of Julia `DataFrames` supporting only one horizontal axis. While multiple horizontal axis DataFrames become hard to analyze and process, they can be useful as the final presentation of the data in some contexts. You can emulate multiple horizontal axis DataFrames unstacking on different columns and then horizontally merging the sub-DataFrames. To obtain the wide format shown in Table 10-3, you could use:

```
wide_df_p = unstack(wide_df,["region","year"],"product",
"production")
wide_df_c = unstack(wide_df,["region","year"],"product",
"consumption")
rename!(wide_df_p, Dict("Hardwood" => "prod_Hardwood",
"Softwood" => "prod_Softwood"))
rename!(wide_df_c, Dict("Hardwood" => "cons_Hardwood",
"Softwood" => "cons_Softwood"))
widewide_df = innerjoin(wide_df_p,wide_df_c,
on=["region","year"] )
```

205

## 10.1.10 The Split-Apply-Combine Strategy

The DataFrames package supports the common *split-apply-combine* strategy through the combine function, which allows you to specify (1) how to split the dataframe into sub-dataframes (called GroupedDataFrame), (2) which operation to perform on each sub-dataframe independently, and (3) how to combine the result of (2) into a final dataframe.

The splitting part is performed by the groupby(df,colNames) function. This returns a sort of iterator on the different groups resulting from the splitting. For example, gdf = groupby(df, ["product"]) would return a GroupedDataFrame with the rows of the original dataframe assigned to two groups based on their value in the "product" column.

While you could manually apply and combine parts of the strategy to these groups (which are technically SubDataFrame), the combine function allows you to do everything in a single pass.

There are two ways to use the combine function. They both work on a GroupedDataFrame.

In the first method, you specify a function and, optionally, a target name for each column you are interested in. This function works independently on each df group. You can also use special functions that don't refer to specific columns, such as nrows (number of rows in the df group) or eachindex (position of the row within the df group). For example, the following snippet calculates the average production and total consumption for each product and region:

```
using Statistics # for the mean function
statistics = combine(groupby(df,["region","product"]),
    "production"  => mean => "avg_prod",
    "consumption" => sum  => "tot_cons",
    nrow
)
```

This results in the following dataframe:

```
4×5 DataFrame
 Row │ region  product   avg_prod  tot_cons  nrow
     │ String  String    Float64   Float64   Int64
─────┼────────────────────────────────────────────
   1 │ US      Hardwood    3.25      11.7      2
   2 │ US      Softwood    2.2       12.3      2
   3 │ EU      Hardwood    2.75       7.5      2
   4 │ EU      Softwood    1.4        9.5      2
```

Note that the columns on which the grouping was based (region and product) are automatically added to the returned dataframe.

The second way to use the combine function is to explicitly pass a function to combine that operates (takes as an argument) on each df group. This function can be given as the first argument or, more conveniently, as a do block.

The previous example becomes

```
statistics2 = combine(groupby(df,["region","product"]))
do subdf
    (avg_prod = mean(subdf.production),
     tot_cons = sum(subdf.consumption),
     nrow     = size(subdf,1))
end
```

The function must return a named tuple in order to be automatically combined, hence the parentheses in the body of the do block. Alternatively, the do block could have returned a horizontal array (a row, and in this case the output dataframe would have the columns automatically named), or it could have returned a dataframe directly:

```
statistics3 = combine(groupby(df,["region","product"]))
do subdf
  [mean(subdf.production) sum(subdf.consumption) size(subdf,1)]
end
statistics4 = combine(groupby(df,["region","product"]))
do subdf
  DataFrame(avg_prod = mean(subdf.production),
  tot_cons = sum(subdf.consumption),
  nrow     = size(subdf,1))
end
```

The preceding examples all perform a *reduction* on the dataframe: several rows are reduced to a single one. Another classic case where you can use the split-apply-combine strategy to return a modified, but not reduced, dataframe is when you want to calculate some cumulative data by group. For example, the following example calculates the cumulative annual timber production by region and product:

```
cumprod = combine(groupby(df,["region","product"])) do subdf
    (year = subdf.year, prod = subdf.production, cumprod =
    cumsum(subdf.production))
end
```
8×5 DataFrame

| Row | region | product | year | prod | cumprod |
|---|---|---|---|---|---|
| | String | String | Int64 | Float64 | Float64 |
| 1 | US | Hardwood | 2010 | 3.3 | 3.3 |
| 2 | US | Hardwood | 2020 | 3.2 | 6.5 |
| 3 | US | Softwood | 2010 | 2.3 | 2.3 |
| 4 | US | Softwood | 2020 | 2.1 | 4.4 |

| 5 | EU | Hardwood | 2010 | 2.7 | 2.7 |
| 6 | EU | Hardwood | 2020 | 2.8 | 5.5 |
| 7 | EU | Softwood | 2010 | 1.5 | 1.5 |
| 8 | EU | Softwood | 2020 | 1.3 | 2.8 |

Alternatively, you can use the split-apply-combine strategy with the `@linq` macro from the `DataFramesMeta.jl` package. The `@linq` macro supports a *Language Integrated Query (LINQ)* style over chained data transformation operations, in a way similar to R's dplyr package (https://cran.r-project.org/web/packages/dplyr/):

```
using DataFramesMeta
cumprod = @linq df                              |>
          groupby([:region,:product])           |>
          transform(:cumprod = cumsum(:production))
```

Here, |> represents the *pipe* operator (covered in depth in a later section), which passes the object returned by the operation on the left as the first argument to the operation on the right: data (df) is first passed to the groupby function to be split, and the split subgroups are finally passed to the transform(df, operation) function to be modified. Note that with the `@linq` macro you have to use symbols to refer to the column names.

## 10.1.11 Dataframe Export

### 10.1.11.1 Exporting to CSV, Excel, or ODS

Exporting DataFrames to CSV, Excel, or ODS format is directly supported by, respectively, the CSV.jl, XLSX.jl, and OdsIO.jl packages, introduced in Chapter 5:

```
CSV.write("file.csv", df)
XLSX.writetable("file.xlsx", df)
OdsIO.ods_write("FILE.ods",Dict(("Sheet1",1,1)=>df))
```

### 10.1.11.2 Exporting to a Matrix

You can use the `Matrix` constructor with the dataframe to convert the dataframe (or a selection of it) into a matrix:

- `Matrix(df)`: Convert to an `Array{Any,2}`.
- `Matrix{Union{Float64, Int64, String}}(df)`: Convert to an `Array{Union{Float64, Int64, String},2}`.
- `Matrix(df[:,["production","consumption"]])`: Pick up homogeneous type columns, resulting in an `Array{Float64,2}`.

### 10.1.11.3 Exporting to a Dict

To convert a `DataFrame` into a dictionary, you can use the following function:

```
function to_dict(df, dim_cols, value_col)
  ktypes = length(dim_cols) == 1 ? eltype(df[!,dim_cols[1]]) :
  Tuple{[eltype(df[!,dc]) for dc in dim_cols]...}
  to_return = Dict{ktypes,eltype(df[!,value_col])}()
  for r in eachrow(df)
    key_values = []
    [push!(key_values,r[d]) for d in dim_cols]
    to_return[(key_values...,)] = r[value_col]
```

```
    end
    return to_return
end
```

`to_dict` converts a dataframe in a dictionary specifying the column(s) to be used as the key (in the given order) and the column to be used to store the value of the dictionary. For example, `to_dict(df,[:region,:product, :year],:production)` will result in the following:

```
Dict{Tuple{String, String, Int64}, Float64} with 8 entries:
  ("US", "Hardwood", 2010) => 3.3
  ("US", "Softwood", 2020) => 2.1
  ("US", "Softwood", 2010) => 2.3
  ("EU", "Hardwood", 2010) => 2.7
  ("EU", "Hardwood", 2020) => 2.8
  ("US", "Hardwood", 2020) => 3.2
  ("EU", "Softwood", 2010) => 1.5
  ("EU", "Softwood", 2020) => 1.3
```

### 10.1.11.4 Exporting to HDF5 Format

The `HDF5.jl` (https://github.com/JuliaIO/HDF5.jl) package doesn't support DataFrames directly, so you'll have to export them as matrices first. Another limitation is that it won't accept a matrix of type `Any` or `Union`, so you may have to export the DataFrame in multiple parts, for example the string and the numeric columns separately:

`h5write("out.h5", "mygroup/df", Matrix(df[:,["production", "consumption"]]))`.

You can read the data back with data = h5read("out.h5", "mygroup/df").

Alternatively, you can use the JLD2.jl (https://github.com/JuliaIO/JLD2.jl) serialization package, which uses the same HDF5 format under the hood, but adds its own specific data attributes:

```
using JLD2
JLD2.jldopen("df.jld", "w") do f
  f["mydf"] = df
end
df2 = JLD2.load("df.jld","mydf")
```

## 10.2 Using IndexedTables

The IndexedTable.jl (https://github.com/JuliaComputing/IndexedTables.jl) package provides alternative table-like data structures. The IndexedTables API is somewhat less convenient than that of DataFrames, but its data structures are particularly interesting (a) for performant row selections, since they internally use indexed named tuples (instead of arrays as in a DataFrame), and (b) because they are part of a large JuliaDB ecosystem (https://juliadb.org/) of data management packages, capable of handling very large datasets that can't be held in the memory of a single machine.

The data structures come in two "flavors": normal indexed tables (IndexedTable) and sparse indexed tables (NDSparse). Of the two, I'll discuss only NDSparse, as IndexedTable is selectable by row position, and is potentially less interesting than NDSparse, where you can use keys instead.

## 10.2.1 Creating an IndexedTable (`NDSParse`)

Sparse indexed tables are created with the `ndsparse` function, which accepts either a single named tuple or a pair of them. The reason is that IndexedTables explicitly splits between *key* columns and *value* columns (the method with a single named tuple assumes the last column to be a single-value column). For example:

```
my_table = ndsparse((
           region      = ["US","US","US","US","EU","EU","EU","EU"],
           product     = ["Hardwood","Hardwood","Softwood","Softwood","Hardwood","Hardwood","Softwood","Softwood"],
           year        = [2020,2010,2020,2010,2020,2010,2020,2010]
        ),(
           production  = [3.3,3.2,2.3,2.1,2.7,2.8,1.5,1.3],
           consumption = [4.3,7.4,2.5,9.8,3.2,4.3,6.5,3.0]
        ))
```

The division between key and value columns is evident when the indexed table is visualized:

```
julia> my_table
3-d NDSparse with 8 values (2 field named tuples):
region   product       year │ production  consumption
─────────────────────────────┼────────────────────────
"EU"     "Hardwood"    2010 │ 2.8         4.3
"EU"     "Hardwood"    2020 │ 2.7         3.2
"EU"     "Softwood"    2010 │ 1.3         3.0
"EU"     "Softwood"    2020 │ 1.5         6.5
"US"     "Hardwood"    2010 │ 3.2         7.4
"US"     "Hardwood"    2020 │ 3.3         4.3
"US"     "Softwood"    2010 │ 2.1         9.8
"US"     "Softwood"    2020 │ 2.3         2.5
```

Further, you may notice that the table has been automatically sorted, like in a dictionary.

Indexed tables can also be obtained, using the same ndsparse function, from an existing dataframe; for example, my_table = ndsparse(df).

(Sparse) indexed tables are accepted as arguments in some self-explaining functions such as show, first, colnames, pkeynames, and keytype.

---

### ❗ Caution

IndexedTables data structures are not suitable for data where the "combined" key (the key across all dimensions) is duplicated. The constructor would still allow you to create the indexed table, but then many methods would throw an error.

---

## 10.2.2 Row Filtering

As previously mentioned, the main advantage of indexed tables is that they allow very fast row selection. Consider this example:

```
t[key_for_dim1,key_for_dim2,:,key_for_dim3,...]
```

This selects the rows that match the given set of keys, with a colon (:) indicating all values for that particular dimension. If the result is a single value, a named tuple is returned; otherwise, another sparse indexed table is returned.

As claimed, the following output shows that the selection is very fast. The example compares the time taken to search for a particular value in the key1 field of an IndexedTable and an equivalent DataFrame.

```
n = 1000000
looked_value = 100
key1 = rand(1:1000,n); key2 = rand(1:1000,n); values = rand(n);
my_table2 = ndsparse((k1 = key1, k2 = key2), (v=values,))
my_df2    = DataFrame(k1 = key1, k2 = key2, v = values)
@benchmark(my_table2[looked_value,:])                   #   24 μs
@benchmark(my_df2[my_df2.k1 .== looked_value,:]) # 1004 μs
```

## 10.2.3 Editing/Adding Values

To change or add values, use the following:

- Change values: my_table["EU","Hardwood",2020] = (consumption = 3.4, production = 2.9)

- Add values: my_table["EU","Hardwood",2012] = (production = 2.8, consumption = 3.3)

That is, the keys must be given as positional arguments within the square brackets, while the values must be given as a named pair (for which the position doesn't matter).

CHAPTER 10   WORKING WITH DATA

## 10.3  Using the Pipe Operator

Chaining (or "piping") allows you to string together multiple function calls in a way that is both compact and readable. It avoids storing intermediate results without having to embed function calls in each other.

With the pipe operator |>, the code to the right of |> operates on the result of the code to the left of it. In practice, the result of the operation on the left becomes the argument of the function call on the right.

Concatenation is very useful for manipulating data. Let's say you want to use the following (silly) functions to operate on some data and print the final result:

```
add6(a) = a+6; div4(a) = a/4;
```

You could either introduce temporary variables or embed the function calls:

```
# Method #1, temporary variables:
a = 2;
b = add6(a);
c = div4(b);
println(c) # Output: 2.0
# Method 2, chained function calls:
println(div4(add6(a)))
```

With piping, you can write the following instead:

```
a |> add6 |> div4 |> println
```

Piping in Julia Base is very limited in the sense that it only supports functions with one argument, and piping over only a single function at a time.

Conversely, the Pipe.jl package provides the @pipe macro, which overrides the |> operator. It allows you to use functions with multiple arguments, where you can use the underscore character (_) as a placeholder for the value of the left side. You can also pipe multiple functions in a single pipe, like this:

---
```
mysum(x,y) = x+y; mydiv(x,y) = x/y
a = 2
# With temporary variables:
b = mysum(a,6)
c = mydiv(4,a)
d = b + c
# With @pipe:
@pipe a |> mysum(_,6) + mydiv(4,_) |> println # Output: 10.0
```
---

You can even pipe over multiple values, embedding them in tuples:

---
```
data = (2,6,4)
# With temporary variables:
b = mysum(data[1],data[2]) # 8
c = mydiv(data[3],data[1]) # 2
d = b + c       # 10
println(d)
# With @pipe:
@pipe data |> mysum(_[1],_[2]) + mydiv(_[3],_[1]) |> println
```
---

Note that, as with the basic pipe operator, functions that require a single argument provided by the piped data don't require parentheses.

CHAPTER 10   WORKING WITH DATA

## 10.4 Plotting

Plotting in Julia can be done using a specific plotting package (e.g., Gadfly.jl (https://github.com/dcjones/Gadfly.jl) or VegaLite.jl (https://github.com/queryverse/VegaLite.jl)) or using a metapackage that provides a wrapper with a unified API to several supported *backends* (e.g., Plots.jl (https://github.com/JuliaPlots/Plots.jl) or Makie.jl (https://github.com/MakieOrg/Makie.jl)).

In this section I will show how to use Plots.jl and some of its backends.

### 10.4.1 Installation and Backends

To use Plots.jl, you need at least one backend. Backends are chosen by calling chosenbackend()—that is, the name of the corresponding backend package but in all lowercase—before calling the plot function. The Plots package comes with GR.jl as the default backend (see Figure 10-1), and GR.jl is installed along with Plots.jl. If you need other backends, just install the appropriate packages.

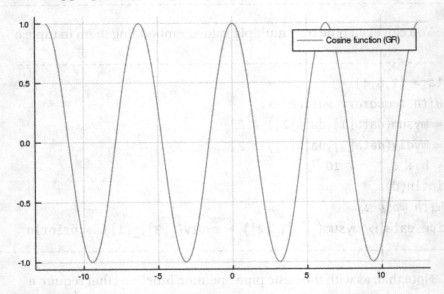

***Figure 10-1.*** *Cosine function with the default backend GR*

218

# CHAPTER 10 WORKING WITH DATA

Two fairly complete backends are PlotlyJS.js (https://github.com/sglyon/PlotlyJS.jl), a Julia wrapper to the Plotly.js (https://plot.ly/) visualization library, and PyPlot.jl (https://github.com/JuliaPy/PyPlot.jl), a wrapper to the Python Matplotlib library (https://matplotlib.org/stable/).

You can use multiple backends in the same session. For example:

```
using Pkg
Pkg.add("Plots")
using Plots
plot(cos,-4pi,4pi, label="Cosine function (GR)") # Plot using
                                                   the default
                                                   GR backend
Pkg.add("PyPlot") # Install the PyPlot backend
pyplot() # Switch backend
plot(cos,-4pi,4pi, label="Cosine function (PyPlot)") # Plot using
                                                       the PyPlot
                                                       backend
Pkg.add("PlotlyJS") # Install the PlotlyJS backend
plotlyjs() # Switch backend
plot(cos,-4pi,4pi, label="Cosine function (PlotlyJS)")
# Plot using the plotlyjs backend
```

While GR is the default backend (i.e., the one that will be used on every new session unless another backend is called), you can set your own default via the following in the ~/.julia/config/startup.jl file:

```
ENV["PLOTS_DEFAULT_BACKEND"] = "my_preferred_backend"
```

## CHAPTER 10 WORKING WITH DATA

> **❶ Important !**
>
> Be careful not to mix different plotting packages (e.g., `Plots` and one of its backends used directly). This will cause problems such as having to prepend calls to the `plot` function (and possibly other functions) with the package name. If you have already imported a plotting package and want to use a different one, you should restart the Julia kernel (this is not necessary when switching between the different backends of the `Plots.jl` package, and is one of its main advantages). You can check which backend is currently used by `Plots` with `backend()`.

There are many different backends. While basic plots can be generated by virtually any backend, each will have its own advantage in some feature, whether it's speed, aesthetics, or interactivity. Some advanced features may be supported only by a subset of backends.

It's important to choose the right backend for your specific needs. The following pages may be of interest:

- Choosing a backend: https://docs.juliaplots.org/stable/backends/
- Charts and attributes supported by various backends: https://docs.juliaplots.org/stable/generated/supported/

## 10.4.2 The `plot` Function

Plots are rendered in the plot panel if you're using VS Code, or in a separate window if you're using the REPL (unless you're using the `UnicodePlots.jl` backend, which uses characters to render the plot in a text terminal).

CHAPTER 10  WORKING WITH DATA

A relatively long first-time-to-plot may still occur, but package loading has improved considerably in recent versions of Julia.

You have already seen an example of one method of the plot function, namely plot(func,l_bound,u_bound;kwargs). Another way is to plot some data using plot(x_data,y_data;kwargs), where x_data is an Array{T,1} of numbers or strings for the coordinates on the horizontal axis and y_data is either another Array{T,1} (i.e., a column vector) or an Array{T,2} (i.e., a matrix). In the first case, a single series is plotted; in the second case, each column is treated as a separate data series (see Figure 10-2):

---

```
x = ["a","b","c","d","e",]
y = rand(5,3)
plot(x,y)
```

---

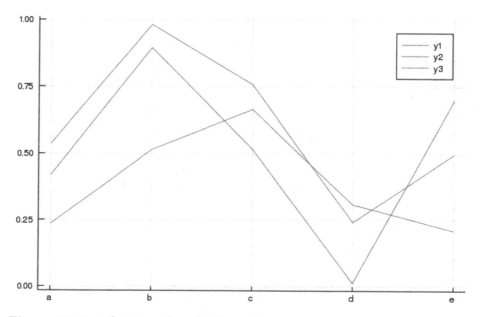

*Figure 10-2. Plotting of multiple series*

While plot objects can be bound to a particular variable like any other object, with my_plot = plot(...), the peculiarity is that the last call to plot defines the "current" plot, which is the default to be visualized or saved. Using the plot! function instead adds elements to an existing plot, the "current" one if not specified otherwise (i.e., by giving the plot object you are operating on as the first argument). An alternative way to plot multiple series is therefore to plot a single series first and then add series one by one with plot!. This snippet produces a plot that is exactly the same as the previous one:

```
p = plot(x,y[:,1])    # create a new "current plot" and assign it
                      to the variable p
plot!(x,y[:,2])       # edit the current plot
plot!(p, x,y[:,3])    # edit the plot assigned to variable p
                      (that is also the current one)
```

The default type of chart drawn is a :line chart. Different types of charts can be specified with the seriestype keyword argument of the plot function, as follows:

```
plot(x,y[:,1]; seriestype=:scatter)
```

Common types of charts are :scatter, :line, and :histogram. You can obtain the full list of supported types with plotattr("seriestype").

> **ℹ Note**
>
> For many type of series, plot defines alias functions, such as scatter(args), which corresponds to plot(args, seriestype=:scatter).

Using first plot and then plot!, you can mix several types of charts together (see Figure 10-3):

```
plot(x,y[:,1], seriestype=:bar)
plot!(x,y[:,2], seriestype=:line)
plot!(x,y[:,3], seriestype=:scatter)
```

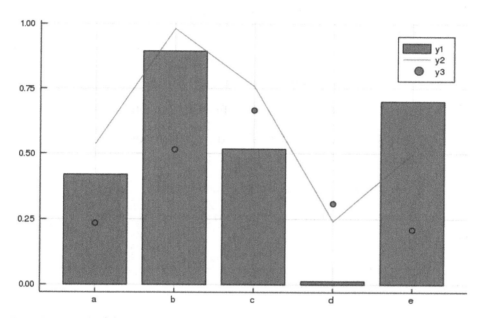

***Figure 10-3.*** *Plotting of different chart types*

seriestype and label are just two of the many keyword arguments supported by plot. They are too numerous to describe in detail in this book. Refer to the Plots.jl documentation or use the inline help in a terminal (?plot and plotattr()) if you want to learn more.

CHAPTER 10  WORKING WITH DATA

## 10.4.3 Plotting from Dataframes

Instead of plotting data from arrays, you can plot data from dataframes directly, using the @df macro provided by the StatsPlots.jl (https://github.com/JuliaPlots/StatsPlots.jl) package (a package that adds statistically oriented functionalities on top of Plots):

```
@df df plot(:year, [:production :consumption], colour = [:red :blue])
```

Using the @df macro has three advantages:

- You don't need to repeat the df name for each series.
- Series are automatically labeled with their column headers.
- You can group data in the plot call itself.

Here is a more elaborate example with grouped series (see Figure 10-4):

```
using DataFrames, StatsPlots
# Let's use a modified version of our example data with more years and just one region:
df = DataFrame(
  product     = ["Softwood","Softwood","Softwood","Softwood",
                 "Hardwood","Hardwood","Hardwood","Hardwood"],
  year        = [2010,2011,2012,2013,2010,2011,2012,2013],
  production  = [120,150,170,160,100,130,165,158],
  consumption = [70,90,100,95,   80,95,110,120]
)
```

224

CHAPTER 10  WORKING WITH DATA

```
mycolours = [:green :orange] # note it's a row vector and the
colors of the series will be alphabetically ordered whatever
order we give it here
timber_plot = @df df plot(:year, :production, group=:product,
linestyle = :solid, linewidth=3, label=reshape(("Production of
" .* sort(unique(:product))) ,(1,:)), color=mycolours)
@df df plot!(:year, :consumption, group=:product, linestyle
= :dot, linewidth=3, label =reshape(("Consumption of " .*
sort(unique(:product))) ,(1,:)), color=mycolours)
```

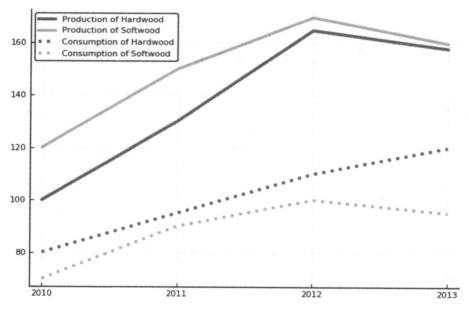

***Figure 10-4.*** *Plotting multiple series from a DataFrame*

The key point of the previous snippet is to use group=:product to split the original series. You then use the color attribute to graphically separate the visualization of these new series. To distinguish between production and consumption, use the linestyle attribute instead.

225

The legend is a little trickier to write because you have to consider two dimensions at the same time. In this example, you hard-code the first dimension (the market variable you are considering, either production or consumption) while concatenating the legend element with the sorted results of unique(:product) for the other dimension. Finally, you reshape it as a row vector, as this is the format expected by the label attribute.

 **Tip**

You don't need to use both StatsPlots and Plots, as the former re-exports the functionality of the latter.

## 10.4.4 Plotting Densities and Distributions

StatsPlots.jl can directly plot densities and analytical distributions from the Distributions.jl (https://github.com/JuliaStats/Distributions.jl) package, which you will examine in detail in Chapter 11.

For example, the following script first creates a distribution object, in this case a normal distribution with a mean of 10 and a standard deviation of 10. It then samples 2000 points from such a distribution. Finally, it plots both the exact probability density function and the empirical density computed from the sampled data (Figure 10-5). The bandwidth parameter can greatly influence the shape of the plotted density; if it is set too small, noise will appear in the plot, as shown in the last plot! call.

```
using Distributions, StatsPlots
dist = Normal(10,10)
data = rand(dist,2000)
plot(dist, label="Exact distribution")
density!(data, label="Emp. dens, def bandwidth")
density!(data, bandwidth=0.5,label="Emp. dens with noises")
```

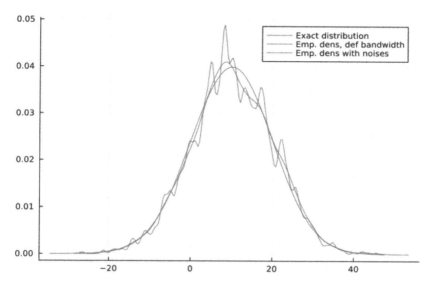

***Figure 10-5.*** *Plotting distributions and densities*

## 10.4.5 Combine Multiple Plots in a Single Figure

To combine several plots into a single figure, you can use the @layout macro, where you specify how the individual plots should be aligned, using placeholders and the semicolon (;) to indicate a new line. You then build the individual plots and assign them to different variables, which you finally use in the plot function together with the layout keyword argument to indicate the layout object you have just created. The following example combines three plots in a single figure—a single plot in the first row and two plots in the second row (see Figure 10-6):

```
l  = @layout [row1 ; r2c1 r2c2];  # create the layout obj
p1 = plot(x, y[:,1]);             # compose 1st plot
p2 = scatter(x, y[:,2]);          # compose 2nd plot
p3 = plot(x, y[:,3]);             # compose 3rd plot
plot(p1, p2, p3, layout = l)      # plot the final figure
```

CHAPTER 10  WORKING WITH DATA

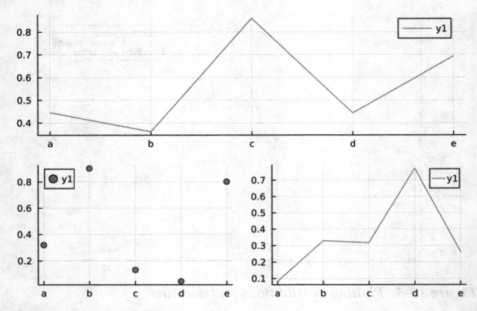

***Figure 10-6.*** *Multi-plots figure*

## 10.4.6 Saving a Plot

To save a plot, call savefig(plot_object,filename), where the extension given to filename determines the format in which the plot will be saved. Some supported formats are .png (standard raster image) .pdf, and .svg (both vector formats). As with the plot function, you can specify the specific plot object to be saved or leave the first argument empty to save the current plot object:

```
savefig(timber_plot, "timber_markets.svg")
savefig("multiple_plots.pdf")
savefig("multiple_plots.png")
```

# CHAPTER 11

# Scientific Libraries

The following third-party packages are covered in this chapter:

| | | |
|---|---|---|
| JuMP.jl | https://github.com/JuliaOpt/JuMP.jl | v1.23.0 |
| HiGHS.jl | https://github.com/ERGO-Code/HiGHS | v1.9.2 |
| Ipopt.jl | https://github.com/JuliaOpt/Ipopt.jl | v1.6.6 |
| SymPyPythonCall.jl | https://github.com/jverzani/SymPyPythonCall.jl | v0.4.0 |
| LsqFit.jl | https://github.com/JuliaNLSolvers/LsqFit.jl | v0.15.0 |
| Distributions.jl | https://github.com/JuliaStats/Distributions.jl | v0.25.110 |

While Julia is a complete, general-purpose programming language, it is undeniable that the focus of its community is currently on mathematical and computational issues. You have already seen in the previous chapters the wide range of libraries for processing numerical data, and how the implications for computational performance have shaped decisions from the very beginning of Julia's development.

This chapter introduces some of the most important libraries for scientific analysis. Some of these, such as JuMP.jl for numerical optimization, are themselves often the main reason for using Julia.

CHAPTER 11   SCIENTIFIC LIBRARIES

Thousands of Julia packages are available. The limited set of packages covered in this chapter focus on relatively general tasks. A few good places to start looking for more specific packages are `https://juliapackages.com/`, `https://juliahub.com/ui/Packages`, and (for a more manually curated list) `https://github.com/svaksha/Julia.jl`.

## 11.1  JuMP, an Optimization Framework

JuMP (`https://github.com/JuliaOpt/JuMP.jl`) is an "Algebraic Modeling Language" (AML) for mathematical optimization problems, similar to GAMS (`https://www.gams.com/`), AMPL (`https://ampl.com/`), and Pyomo (`http://www.pyomo.org/`).

Optimization problems are problems that find the "optimal" values of decision variables in the context of an objective function and the constraints between the variables. Constraints occur everywhere in the real world, be it in business decisions (e.g., "How much of X should we produce?"), engineering settings (e.g., "At what angle should we design a flap for maximum efficiency?"), or personal choices (e.g., "A small apartment in the city or a large house in the suburbs?").

Optimization problems can be formulated mathematically in terms of a function to maximize (or minimize) subject to other functions acting as constraints. Software "tools" have been developed to solve these problems. They typically require you to formulate the optimization problem in matrix form and, for nonlinear problems, to provide some additional information about the problem in the form of first and second derivatives to help find the numerically optimal solution. Solver-based approaches in the MATLAB optimization toolbox or R's `optim` function are two well-known examples.

AMLs take a different approach in that they decouple the formulation of the problem from the specific algorithm used to solve it. The problem then can be algebraically encoded, using the same mathematical notation that is used to describe it.

Crucially, variables can be indexed in any dimension using the concept of "sets," allowing you to write equations such as this:

```
Demand[year, time, region, product] = Supply[year, time, region, product]
```

Tools to automatically provide the derivatives are included.

Once the problems are formulated, they can be solved by any "solver engine" that is supported by the AML software and suitable for that particular problem class.

However, a disadvantage of commercial AMLs is that they each use their own specific language. Being language specific, they lag behind "real" general-purpose programming languages in many features (data handling, availability of developer tools, language constructs, etc.). Interfacing them with other (non-optimization) parts of a larger model could be challenging.

By contrast, JuMP is part of a "third generation" of optimization tools (the first being the solver engines themselves and the second being the commercial, language-specific AMLs), which are AMLs embedded as a software library in a general-purpose programming language.

JuMP offers the same ease of modeling and solver independence of language-specific AMLs, while also providing the benefits of a much more practical modeling environment. When combined with free solver engines (e.g., HiGHS for mixed integer problems and IPOPT for nonlinear problems), JuMP is a complete open source solution.

In this section you will learn how to use JuMP by implementing two problems, one with only linear functions and one where the objective is nonlinear. The general settings of the problems will be the same: you choose a suitable solver for the specific problem, you load the Julia package for the solver-specific interface, you create the problem "object," and you add to it its characteristic elements, such as variables, objective functions, and constraints. At this point you can "solve" the problem and retrieve its solutions.

CHAPTER 11   SCIENTIFIC LIBRARIES

The examples also serve to demonstrate the power of Julia macros. Using them, JuMP implements its own language while still remaining Julia.

## 11.1.1  The Transport Problem: A Linear Problem
### 11.1.1.1  The Problem

This is the implementation in JuMP of the basic transport model used in the GAMS tutorial.

Given a single product, $p$ plants and $m$ markets, the problem is to define the best routes between $p$ and $m$ to minimize transport costs while respecting the capacity limits ($a_p$) of each plant and satisfying the demand of each market ($b_m$):

$$\min \sum_p \sum_m c_{p,m} * x_{p,m}$$

Subject to:

$$\sum_m x_{p,m} \leq a_p$$

$$\sum_p x_{p,m} \geq b_m$$

where $c_{p,m}$ are the unit transport costs between $p$ and $m$ and $x_{p,m}$ are the transported quantities (to be chosen). The original formulation is from *Linear Programming and Extensions* (Chapter 3.3) by George Dantzig (Princeton University Press: 1963), while its GAMS implementation can be found in *GAMS: A User's Guide* (Chapter 2) by Richard Rosenthal (The Scientific Press: 1988) or at this link: https://www.gams.com/mccarl/trnsport.gms.

To facilitate comparison, I leave the equivalent GAMS code for each operation in the comments.

## 11.1.1.2 Importing the Libraries

First you need to import the `JuMP.jl` package. Then you need the package that implements the specific interface with the desired solver engine. In this case, the problem is a linear problem, so you will use the HiGHS library and install the `HiGHS.jl` Julia package.

---

 **Note**

Both the `HiGHS.jl` and `Ipopt.jl` packages will automatically install the necessary binaries for their own use. If you want to use custom builds of the solver engines, see the tutorial at https://jump.dev/JuMP.jl/stable/developers/custom_solver_binaries/.

---

The CSV package is required to load the data:

```
using CSV, JuMP, HiGHS
```

## 11.1.1.3 Defining the Sets

JuMP doesn't have its own syntax for sets, but it uses the native containers available in the core Julia language. Variables, parameters, and constraints can be indexed using these containers. While you can work with position-based lists, it is better to use dictionaries instead (with a minor trade-off in efficiency). So the "sets" in these examples are represented as lists, but then everything else (variables, constraints, and parameters) is a dictionary with the elements of the list as keys.

CHAPTER 11  SCIENTIFIC LIBRARIES

```
# Define sets #
#   Sets
#       i   canning plants    / seattle, san-diego /
#       j   markets           / new-york, chicago, topeka / ;
plants  = ["seattle","san-diego"]          # canning plants
markets = ["new-york","chicago","topeka"]  # markets
```

## 11.1.1.4 Defining the Parameters

The capacity of plants and the demand of markets are defined directly as dictionaries, while the distance is first read as a DataFrame from a whitespace-separated table. It is then converted into a "(plant, market) ⇒ value" dictionary.

```
# Define parameters #
#   Parameters
#       a(i)  capacity of plant i in cases
#         /   seattle      350
#             san-diego    600 /
a = Dict(                  # capacity of plant i in cases
  "seattle"    => 350,
  "san-diego"  => 600,
)
#       b(j)  demand at market j in cases
#         /   new-york     325
#             chicago      300
#             topeka       275 / ;
b = Dict(                  # demand at market j in cases
  "new-york"   => 325,
  "chicago"    => 300,
```

234

```
    "topeka"    => 275,
)

# Table d(i,j)   distance in thousands of miles
#                    new-york       chicago       topeka
#     seattle        2.5            1.7           1.8
#     san-diego      2.5            1.8           1.4    ;
d_table = CSV.read(IOBuffer("""
plants     new-york  chicago  topeka
seattle    2.5       1.7      1.8
san-diego  2.5       1.8      1.4
"""),DataFrame, delim=" ", ignorerepeated=true)
d = Dict( (r[:plants],m) => r[Symbol(m)] for r in eachrow
(d_table), m in markets)
# Here we are converting the table in a "(plant, market) =>
distance" dictionary
# r[:plants]:   the first key, using the cell at the given row
and `plants` field
# m:            the second key
# r[Symbol(m)]: the value, using the cell at the given row and
the `m` field

# Scalar f  freight in dollars per case per thousand
miles  /90/ ;
f = 90 # freight in dollars per case per thousand miles

# Parameter c(i,j)  transport cost in thousands of dollars
per case ;
#              c(i,j) = f * d(i,j) / 1000 ;
# We first declare an empty dictionary and then we fill it with
the values
c = Dict() # transport cost in thousands of dollars per case ;
[ c[p,m] = f * d[p,m] / 1000 for p in plants, m in markets]
----------------------------------------------------------------
```

## 11.1.1.5 Declaring the Model

In this step, you declare a JuMP optimization model and give it a name. This name will be passed as the first argument to all subsequent operations, such as creating variables, constraints, and objective functions (you can work with several models at the same time if you prefer).

You can specify the solver engine at this point in the Model constructor (as is done here) or, equivalently, at a later time (but before solving the problem) with a call to set_optimizer(model,optimizer). You can also specify solver engine-specific options.

```
# Model constructor (transport model)
trmodel = Model(HiGHS.Optimizer) # we choose HiGHS as the
solver engine
set_optimizer_attribute(trmodel, "parallel", "on")
set_optimizer_attribute(trmodel, "output_flag", true)
```

## 11.1.1.6 Declaring the Model Variables

Variables are declared and defined in the @variable macro and can have multiple dimensions—that is, they can be indexed under multiple indexes. Bounds are given at the same time as they are declared. Unlike with GAMS, you don't need to define the variable that is on the left side of the objective function.

```
## Define variables ##
# Variables
#     x(i,j)   shipment quantities in cases
#     z        total transportation costs in thousands of
dollars ;
```

```
# Positive Variable x ;
@variables trmodel begin
    x[p in plants, m in markets] >= 0 # shipment quantities
                                        in cases
end
```

---

### 11.1.1.7 Declaring the Model Constraints

As in GAMS, each constraint can actually be a "family" of constraints, as it can be indexed over the defined sets:

---

```
## Define constraints ##
# supply(i)   observe supply limit at plant i
# supply(i) .. sum (j, x(i,j)) =l= a(i)
# demand(j)   satisfy demand at market j ;
# demand(j) .. sum(i, x(i,j)) =g= b(j);
@constraints trmodel begin
    supply[p in plants],   # observe supply limit at plant p
        sum(x[p,m] for m in markets)  <=  a[p]
    demand[m in markets],  # satisfy demand at market m
        sum(x[p,m] for p in plants)   >=  b[m]
end
```

---

### 11.1.1.8 Declaring the Model Objective

Unlike constraints and variables, the objective is a unique, single function. Note that this is where you specify the direction of the optimization.
`# Objective`

```
@objective trmodel Min begin
    sum(c[p,m]*x[p,m] for p in plants, m in markets)
end
```

### 11.1.1.9 Human-Readable Visualization of the Model (Optional)

If you want, you can use the following command to get a printout of the optimization model in human-readable form, enabling you to check that it is correct:

```
print(trmodel) # The model in mathematical terms is printed
```

### 11.1.1.10 Resolution of the Model

At this point, you can use the following function to call the solver and pass the model to the solver engine for solution. Note the exclamation mark indicating that the trmodel object will be modified by the call.

```
optimize!(trmodel)
```

 **Tip**

You can solve a model with multiple solver engines by pairing the model to a different solver engine with set_optimizer(model, solver) and calling optimize! each time.

### 11.1.1.11 Visualization of the Results

You can check the status returned by the solver with termination_status(model_object) (hoping for an MOI.OPTIMAL) and retrieve the values of the objective function and variables with objective_value(model_object) and value(variable_name), respectively. Duals can be obtained by calling dual(constraint_name). You can then print/visualize them as you prefer. For example:

```
status = termination_status(trmodel)
if (status == MOI.OPTIMAL || status == MOI.LOCALLY_SOLVED ||
status == MOI.TIME_LIMIT) && has_values(trmodel)
    if (status == MOI.OPTIMAL)
        println("** Problem solved correctly **")
    else
        println("** Problem returned a (possibly suboptimal)
        solution **")
    end
    println("- Objective value (total costs): ",
    objective_value(trmodel))
    println("- Optimal routes:")
    optRoutes = value. (x)
    [println("$p --> $m: $(optRoutes[p,m])") for m in markets,
    p in plants]
    println("- Dual of supply:")
    [println("$p = $(dual(supply[p]))") for p in plants]
    println("- Dual of demand:")
    [println("$m = $(dual(demand[m]))") for m in markets]
else
```

```
    println("The model was not solved correctly.")
    println(status)
end
```

This should print as follows:

```
** Problem solved correctly **
- Objective value (total costs): 153.675
- Optimal routes:
seattle --> new-york: 0.0
seattle --> chicago: 300.0
seattle --> topeka: 0.0
san-diego --> new-york: 325.0
san-diego --> chicago: 0.0
san-diego --> topeka: 275.0
- Dual of supply:
seattle = 0.0
san-diego = 0.0
- Dual of demand:
new-york = 0.225
chicago = 0.153
topeka = 0.12599999999999997
```

> **! Caution**
>
> The reported dual values are positive because a marginal increase in demand in market m of "x" leads to the corresponding dual value by x increase in transport costs (e.g., setting demand in Chicago to 301 units leads to transport costs of 153.828). Use shadow_price instead of dual to get the economic interpretation with the correct sign depending on the direction of the optimization and the constraint.

## 11.1.2 Choosing Between Pizzas and Sandwiches: A Nonlinear Problem

The only difference with nonlinear problems is that you have to provide good starting points for the variables (try running the following example with 0 as the starting point!). Note that you don't need to provide any other information (in particular, no derivatives are needed).

### 11.1.2.1 The Problem

A student has to decide how to allocate her weekly food budget of $80. She can choose only between pizzas (p) and sandwiches (s). Pizzas cost $10, while sandwiches cost only $4.

Her "utility function" (how happy she is with a given combination of pizzas and sandwiches) is ($100p - 2p^2 + 70s - 2s^2 - 3ps$). She likes both, with a preference for pizza.

This is a simple problem of maximizing a nonlinear function of two variables subject to a single constraint. To keep the example simple, we don't use sets here (i.e., both variables and constraints are just scalars):

$$max_{p,s} \, 100p - 2p^2 + 70s - 2s^2 - 3ps$$

subject to

$$10p + 4s \leq 80$$

The first equation is the utility function, and the second equation is what economists call the "budget constraint" (i.e., we can't spend more money than we have).

## 11.1.2.2 Importing the Libraries and Declaring the Model

Since you're dealing with a nonlinear model, you need to choose a solver engine capable of handling this type of problem. IPOPT is a good choice:

```
using JuMP, Ipopt
m = Model()
set_optimizer(m,Ipopt.Optimizer)
set_optimizer_attribute(m, "print_level", 0)
```

## 11.1.2.3 Declaring Model Variables, Constraints, and Objective

Since these are physical quantities, in addition to the explicit constraint, you need to add a lower bound of 0 to your variables (lower and upper bounds can be specified in an expression as `lB <= var <= uB`).

```
@variable(m, 0 <= p, start=1, base_name="Quantities of pizzas")
@variable(m, 0 <= s, start=1, base_name="Quantities of sandwiches")
@constraint(m, budget,    10p + 4s <= 80 )
@objective(m, Max, 100*p - 2*p^2 + 70*s - 2*s^2 - 3*p*s)
```

You can interpret the problem as the optimal "weekly average," so it is okay if the optimal result is half a pizza. Otherwise, you should restrict the variables to integers by adding the `integer=true` option to their declaration. But then the class of the problem would change to Mixed-Integer Nonlinear Programming (MINLP) and you would have to change the solver engine as well (Bonmin is good for solving MINLP problems).

## 11.1.2.4 Resolving the Model and Visualizing the Results

Because the problem is nonlinear, the solution is reported as a "local optima." It is then up to you to decide whether the nature of the problem guarantees that the local optima is also a global one (as in this case).

---

```
optimize! (m)
status = termination_status(m)
if (status == MOI.OPTIMAL || status == MOI.LOCALLY_SOLVED ||
status == MOI.TIME_LIMIT) && has_values(m)
    if (status == MOI.OPTIMAL)
        println("** Problem solved correctly **")
    elseif (status == MOI.LOCALLY_SOLVED)
        println("** Problem returned a (possibly local) optimal
        solution **")
    else
        println("** Problem returned a (possibly suboptimal)
        solution **")
    end
    println("- Objective value : ", objective_value(m))
    println("- Optimal solutions:")
    println("   - pizzas: $(value. (p))")
    println("   - sandwiches: $(value. (s))")
    println("- Shadow price (budget): $(shadow_price.(budget))")
else
    println("The model was not solved correctly.")
    println(status)
end
```

---

The following is the expected output:

```
** Problem returned a (possibly local) optimal solution **
- Objective value : 750.892861640351
- Optimal solutions:
  - pizzas: 4.642857242786692
  - sandwiches: 8.392857092921894
- Shadow price (budget): 5.6249999750627255
```

## 11.2 SymPy, a CAS System

Another useful mathematical package is SymPyPythonCall.jl (https://github.com/jverzani/SymPyPythonCall.jl), a wrapper for the Python library SymPy (https://www.sympy.org/en/index.html) for symbolic computation, which is the *analytic* resolution of derivatives, integrals, equations (or systems of equations), and so on.

SymPy is a very large library, a *computer algebra system*, with features ranging from basic symbolic arithmetic to quantum physics. This section presents an application of the library that solves the same problem as in the previous section, but this time analytically. See the SymPy documentation (https://docs.sympy.org/latest/index.html) for the complete API (much of it has a direct SymPyPythonCall.jl counterpart).

 **Note**

At the time of writing, there are two identical packages that wrap the SymPy Python library: the older SymPy.jl and the newer SymPyPythonCall.jl. They differ in the inner component used to interface with Python, but are otherwise identical. I am using SymPyPythonCall.jl here. If you have problems, just install

SymPy.jl and replace using SymPyPythonCall with using SymPy. Other Julia symbolic computation packages (Symbolics.jl (https://symbolics.juliasymbolics.org/), SymEngine.jl (https://github.com/symengine/symengine.jl), etc.) may be more computationally efficient, but have a limited subset of functionality.

## 11.2.1 Loading the Library and Declaring Symbols

Variables can be declared using the @symp x y z syntax. You can also declare the assumptions related to the domain of the variables, such as positive, real, integer, odd, and so on (the full list is available in the SymPy documentation (https://docs.sympy.org/dev/modules/core.html#module-sympy.core.assumptions)), either by annotating the variable (e.g., @syms α::positive) or, more explicitly, by using symbols("symbol_name"; assumptions) (e.g., α = symbols("α", integer=true, positive=true)).

You can solve the same example of choosing between pizzas ($q_p$) and sandwiches ($q_s$), but here keeping the price of the pizza ($p_p$) and that of the sandwich ($p_s$) symbolic, just to show that the solution obtained is analytic in nature. It is only later that you will get a specific numerical value. I simplify the example by imposing that the entire budget must be used. This allows you to use the Lagrangian multipliers method to find the analytic solution. (If this is new to you, don't worry. Just follow how SymPy is used to manipulate symbolic equations and find the numerical solution later, when symbols in the analytical solution are replaced by actual numerical values).

---
**using** SympyPythonCall
@syms $q_p$::positive $q_s$::positive $p_p$::positive $p_s$::positive λ

---

## 11.2.2 Creating and Manipulating Expressions

After you have declared the symbols, you can use them to create algebraic expressions (functions). Note that $q_p$, $q_s$, and so forth are now both Julia variables and SymPy.jl "symbols," a reference to the underlying SymPy structure. The same applies to the expressions you create (not to be confused with Julia language expressions):

---

```
julia> typeof(:(q_p+q_s))  # A Julia language Expression
                   Expr
julia> typeof(:(q_p))      # A Julia Symbol
Symbol
julia> typeof(q_p+q_s)     # A SymPy expression, i.e. a
                             SymPy object
Sym{PythonCall.Core.Py}
```

---

You can create the so-called "Lagrangian" by adding to the objective solution each constraint multiplied by the corresponding "Lagrangian multiplier." You can then obtain the partial derivatives of this Lagrangian function using `diff(function,variable)`:

---

```
utility = 100*q_p - 2*q_p^2 + 70*q_s - 2*q_s^2 - 3*q_p*q_s
budget  = p_p* q_p + p_s*q_s
lagr    = utility + λ*(80 - budget)
dlq_p   = diff(lagr,q_p)
dlq_s   = diff(lagr,q_s)
dlλ     = diff(lagr,λ)
```

---

As claimed, each derivative is expressed in symbolic terms:

```
julia> dlqₚ
-pₚ·λ - 4·qₚ - 3·qₛ + 100
```

## 11.2.3 Solving a System of Equations

The first-order conditions tell you that the solutions, expressed in terms of $q_p$, $q_s$, and $\lambda$, are found by setting the relative derivatives of the Lagrangian equal to zero. In other words, you are solving a system of equations with three variables and three unknowns.

solve((equations),(variables)) does exactly that:

```
julia> sol = solve((Eq(dlqₚ,0), Eq(dlqₛ,0), Eq(dlλ,0)),(qₚ, qₛ,
       λ)) # SymPy.solve if a function named "solve" has been already
            defined in other packages
Dict{Sym{PythonCall.Core.Py}, Sym{PythonCall.Core.Py}} with 3
entries:
  λ  => (95*pₚ - 10*pₛ - 280)/(2*pₚ^2 - 3*pₚ*pₛ + 2*pₛ^2)
  qₚ => (-35*pₚ*pₛ + 160*pₚ + 50*pₛ^2 - 120*pₛ)/(2*pₚ^2 - 3*pₚ*pₛ
       + 2*pₛ^2)
  qₛ => (35*pₚ^2 - 50*pₚ*pₛ - 120*pₚ + 160*pₛ)/(2*pₚ^2 - 3*pₚ*pₛ
       + 2*pₛ^2)
```

### 💡 Tip

The solve function accepts expressions directly and sets them to zero, so this would have worked too: solve((dlqₚ, dlqₛ, dlλ),(qₚ, qₛ, λ))).

## 11.2.4 Retrieving Numerical Values

You can use my_expression.evalf(subs=Dict(symbol=>value)) to "inject" the given numerical values of its symbol into the expression and retrieve the corresponding numerical value of the expression.

To retrieve the numerical values of the solutions, you provide the prices of pizzas as $10 and sandwiches as $4:

---
```
q_p_num = sol[q_p].evalf(subs=Dict(p_p=>10,p_s=>4))
# 4.64285714285714
q_s_num = sol[q_s].evalf(subs=Dict(p_p=>10,p_s=>4))
# 8.39285714285714
λ_num = sol[λ].evalf(subs=Dict(p_p=>10,p_s=>4))    # 5.625
```
---

> **ⓘ Note**
>
> Even when visualized as numbers, $q_p$_num, $q_s$_num, and λ_num remain SymPy objects. To convert them to Julia numbers, you can use N(SymPySymbol). For example, N($q_p$_num).

Finally, to get the numerical value of the objective function, replace the symbols in the utility function with these numerical values:

---
```
z = utility.evalf(subs=Dict(q_p=>q_p_num, q_s=>q_s_num))
#750.892857142857
```
---

Congratulations! You have obtained the same "optimal" values that you obtained numerically with JuMP (as expected).

## 11.3 LsqFit, a Data Fit Library

As a third example of scientific libraries, I have chosen a data-fitting model. Statistics (and the related field of machine learning) is probably the area with the most active development in the Julia community. The LsqFit.jl package offers a lot of flexibility while remaining easy to use. You will use it to estimate the logistic growth curve (Verhulst model) of a forest stand of beech (a broad-leaved species), given a sample of timber volume data from a stand in north-west France[1].

### 11.3.1 Loading the Libraries and Defining the Model

You start by specifying the functional form of the model you want to fit and where the parameters to be estimated are located. The @. macro is a useful tool to avoid having to specify the point operator (for broadcasting) for each operation.

The model you want to estimate is the well-known logistic function:

```
using LsqFit,CSV,DataFrames,Plots
# **** Fit volumes to logistic model ***
@. model(age, pars) = pars[1] / (1+exp(-pars[2] * (age - pars[3])) )
obsVols = CSV.read(IOBuffer("""
age    vol
20     56
35     140
```

---

[1] Data from Ecole Nationale du Génie Rural, des Eaux et des Forêts (ENGREF), *Tables de production pour les forets francaises*, 2e édition, p. 80 (1984). Age is given in years, volumes in metric cubes per hectare.

```
60    301
90    420
145   502
"""),DataFrame,delim=" ", ignorerepeated=true)
```

## 11.3.2 Parameters

As the data-fitting algorithm is nonlinear, you will need to provide some "reasonable" starting points. In the case of logistic functions, the first parameter is the maximum level reached by the function (the "carrying capacity"), the second is the growth rate, and the third is the midpoint on the x scale. The following are reasonable starting points:

```
par0 = [600, 0.02, 60]
```

## 11.3.3 Fitting the Model

After have defined the model, you can fit it with your data, your initial guess of the parameters, and, optionally, their lower and/or upper bounds. This is the least-squares fitting stage.

```
par_lb = [ 200, 0.001,    5]
par_ub = [2000, 0.1  , 1000]
fit = curve_fit(model, obsVols.age, obsVols.vol,
par0,lower=par_lb, upper=par_ub)
```

## 11.3.4 Retrieving the Parameters and Comparing the Results with the Observations

The fitted parameters are now available in the fit.param array. At this point you can simply use them to calculate the fitted volumes over each year and compare them with the observed values (see Figure 11-1):

```
fit.param # [497.07, 0.05, 53.5]
fitX = 0:maximum(obsVols.age)*1.2
fitY  = [fit.param[1] / (1+exp(-fit.param[2] * (y - fit.param[3]) ) ) for y in fitX]
plot(fitX,fitY, seriestype=:line, label="Fitted values")
plot!(obsVols.age, obsVols.vol, seriestype=:scatter, label="Obs values")
plot!(obsVols.age, fit.resid, seriestype=:bar, label="Residuals")
```

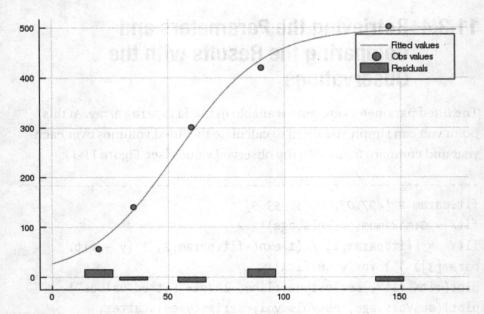

***Figure 11-1.***  *Observed data versus fitted data*

## 11.4 Working with Distributions

A backbone of many statistical and machine learning packages is Distributions.jl (https://github.com/JuliaStats/Distributions.jl), which provides the necessary tools to work with probabilistic analysis.

The syntax is very elegant. First, you define a specific distribution object together with its parameters; for example, Uniform(a,b) for a continuous uniform distribution in [a,b], or Geometric(p) for a geometric distribution with probability p. Then, using a common API (described later in this section), you extract the information of interest from this distribution, or you sample from it.

## 11.4.1 Main Supported Distributions

This section presents a subset of the distributions supported by the Distributions.jl (https://github.com/JuliaStats/Distributions.jl) package. See its documentation at https://juliastats.org/Distributions.jl/stable/ for the full list, or check out my own cheat sheet at https://github.com/sylvaticus/commonDistributionsInJuliaPythonR for their interpretation and a nice comparison with the equivalent R and Python packages.

### Discrete distributions

| Constructor | Parameters |
| --- | --- |
| DiscreteUniform(lRange,uRange) | lRange, uRange $\in \mathbb{Z}$ with uRange $\geqq$ lRange |
| Bernoulli(p) | $p \in [0,1]$ |
| Binomial(n,p) | $n \in \mathbb{N}, p \in [0,1]$ |
| Categorical($p_1$, $p_2$, ..., $p_k$) | $p_1, p_2, ..., p_k$ with $p_k \in [0,1]$, $\sum_{k=1}^{K} p_k = 1$ |
| Multinomial(n, $p_1$, $p_2$, ..., $p_k$) | $n, p_1, p_2, ..., p_k$ with $p_k \in [0,1]$, $\sum_{k=1}^{K} p_k = 1$ and $n \in \mathbb{N}$ |
| Geometric(p) | $p \in [0,1]$ |
| Hypergeometric(nS, nF, nTrials) | nS, nF, nTrials $\in \mathbb{N}_0$ |
| Poisson(rate) | rate $\in \mathbb{R}^+$ |
| NegativeBinomial(nSucc,p) | nSucc $\in \mathbb{N}, p \in [0,1]$ |

## Continuous distributions

| Constructor | Parameters |
| --- | --- |
| Uniform(lRange,uRange) | lRange, uRange $\in \mathbb{R}$ with uRange $\geq$ lRange |
| Exponential(rate) | rate $\in \mathbb{R}^+$ |
| Laplace(loc, scale) | loc $\in \mathbb{R}$, scale $\in \mathbb{R}^+$ |
| Normal(mean,sd) | mean $\in \mathbb{R}$, sd $\in \mathbb{R}^+$ |
| MVNormal(means,cov_matrix) | means $\in \mathbb{R}^d$, cov_matrix $\in \mathbb{R}^{d \times d}$ |
| Erlang(n,rate) | n $\in \mathbb{N}$, rate $\in \mathbb{R}^+$ |
| Cauchy(loc, scale) | loc $\in \mathbb{R}$, scale $\in \mathbb{R}^+$ |
| Chisq(df) | df $\in \mathbb{N}$ |
| TDist(df) | df $\in \mathbb{R}^+$ |
| FDist(df$_1$, df$_2$) | df$_1 \in \mathbb{N}$ df$_2 \in \mathbb{N}$ |
| Beta(shape$\alpha$,shape$\beta$) | shape$\alpha$, shape$\beta \in \mathbb{R}^+$ |
| Gamma(shape$\alpha$,1/rate$\beta$) | shape$\alpha \in \mathbb{R}^+$, rate$\beta \in \mathbb{R}^+$ |

 **Caution**

Note that while the unidimensional normal distribution is parametrized with the standard deviation, the multinomial normal distribution is parametrized with the covariance matrix.

CHAPTER 11  SCIENTIFIC LIBRARIES

## 11.4.2 API

Once you have a distribution object d, you can

- Get its specific parameters with params(d)

- Calculate its mean, variance, and median with mean(d), var(d), and median(d), respectively

- Sample from it with rand(d). Sampling is perhaps one of the best examples of composability in Julia. You can sample an array of random values from d using a custom RNG with rand(RNG,d,array_shape).

- Compute the probability of the random variable taking value x or the probability density at x with pdf(d,x).

- Return the cumulative distribution function at x with cdf(d,x) and its inverse (quantile) with quantile(d,y).

- Plot the probabilities of discrete distributions with bar(x -> pdf(d,x),domain_range) and for continuous distributions with plot(d).

For example, for a binomial (a discrete distribution) with 100 trials and a 20% probability of success on each trial, you have the following (see Figure 11-2):

```
using Distributions, StatsPlots
d = Binomial(100,0.2) # number of trials, prob single trial
params(d) # (100, 0.2)
mean(d)   # 20.0
var(d)    # 16.0
median(d) # 20
rand(d)   # 13
using Random
```

## CHAPTER 11  SCIENTIFIC LIBRARIES

```
rand(Xoshiro(123),d,(2,3)) # [31 18 23; 16 16 19]
pdf(d,20)              # 0.099
y = cdf(d,25)          # 0.91
quantile(d,0.91) # 25
bar(x -> pdf(d,x),1:40,label="Binomial(n=100,p=0.2)")
```

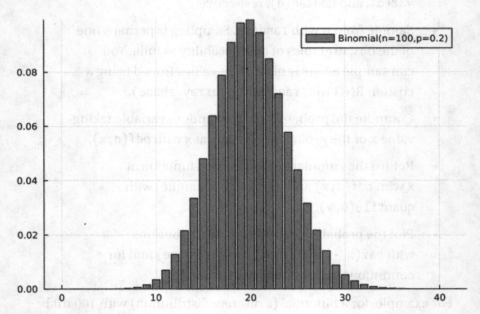

*Figure 11-2.  Plot of a discrete distribution*

## 11.5  EXERCISE 2: Fitting a Forest Growth Model

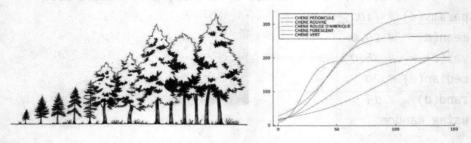

How fast do forests grow? How much wood can they produce per hectare? Forests provide many ecosystem services, but in this simplified exercise you will focus on these two questions by looking at the so-called "raw data" provided by the French National Forest Inventory[2], both in terms of individual trees and in terms of inventoried plots, in order to fit a generic growth model of forest stands in terms of volume in relation to the age of the trees.

**Skills used**: downloading and importing data from the Internet; manipulating tabular data: calculate new columns (fields) based on existing ones, filter rows (records), join tables; fitting a generic model (curve) to data using `LsqFit.jl`.

## 11.5.1 Instructions

The skeleton of the exercise presented next includes some code that is already set up, which you must complete by replacing [...] Write your code here with your own code. If you want to avoid typing, you can find this skeleton in the GitHub repository of the book (https://github.com/Apress/Julia-Quick-Syntax-Reference-2nd-ed), where you will also find its solution.

## 11.5.2 Skeleton

### 11.5.2.1 STEP 1: Set up the environment

First, set the working directory to the directory where this file (ch11-ScientificLibraries.jl if you are using the skeleton provided) is located, and enable it. If you have the provided Manifest.toml file in the directory, just run Pkg.instantiate(); otherwise, add the packages Pipe.jl, HTTP.jl, CSV.jl, DataFrames.jl, LsqFit.jl, and StatsPlots.jl manually. Also fix the random seed for reproducibility.

---

[2] IGN – Inventaire forestier national français, Données brutes, Campagnes annuelles 2005 et suivantes, https://inventaire-forestier.ign.fr/dataIFN/

```
# [...] Write your code here
```

### 11.5.2.2 STEP 2: Load the packages

Load the packages Pipe.jl, HTTP.jl, CSV.jl, DataFrames.jl, LsqFit.jl, and StatsPlots.jl.

```
# [...] Write your code here
```

### 11.5.2.3 STEP 3: Load the data

Load from the Internet the following datasets:

```
ltURL     = "https://bit.ly/apress_julia_alive_trees"
            # alive individual trees data
dtURL     = "https://bit.ly/apress_julia_dead_trees"
            # dead individual trees data
pointsURL = "https://bit.ly/apress_julia_inv_points"
            # plot level data
docURL    = "https://bit.ly/apress_julia_inv_doc"
            # optional, needed for the species label
```

You can make for each of the datasets a @pipe macro starting with HTTP.get(URL).body, continuing the pipe with CSV.File(_), and ending the pipe with a DataFrame object.

```
# [...] Write your code here
```

# CHAPTER 11 SCIENTIFIC LIBRARIES

## 11.5.2.4 STEP 4: Filter out unused information

These datasets have many variables that you will not be using in this exercise. Of all the variables, select only the columns idp (pixel id), c13 (circumference at 1.30 meters), and v (volume of the tree) for the lt and dt dataframes. The two datasets are then concatenated vertically into a total trees dataset. For the Points dataset, select only the variables idp (pixel id), esspre (main forest species code of the stand) and cac (age class).

```
# [...] Write your code here
```

## 11.5.2.5 STEP 5: Compute the timber volumes per hectare

The French National Forest Inventory is based on a concentric sample method: small trees are sampled only over a small area (6 meters radius), intermediate trees are sampled on a concentric area of 9 meters, and large trees (with a circumference larger than 117.5 cm) are sampled on a concentric area of 15 meters radius. Thus, define the following function to compute the contribution of each tree to the volume per hectare:

```
"""
    vHaContribution(volume,circumference)

Return the contribution in terms of m³/ha of the tree.

The French inventory system is based on a concentric sample
method: small trees are sampled on a small area (6 meters
radius), intermediate trees on a concentric area of 9 meters
and only large trees (with a circumference larger than 117.5
cm) are sampled on a concentric area of 15 meters of radius.
```

This function normalizes the contribution of each tree to m³/ha.
"""
```
function vHaContribution(v,c13)
    if c13 < 70.5
        return v/(6^2*pi/(100*100))
    elseif c13 < 117.5
        return v/(9^2*pi/(100*100))
    else
        return v/(15^2*pi/(100*100))
    end
end
```

---

Use the preceding function to compute trees.vHa (the contribution to volume per hectare of each individual tree) based on trees.v (the estimated log volume of each tree) and trees.c13 (the measured circumference of the trees).

---
```
# [...] Write your code here
```
---

## 11.5.2.6 STEP 6: Aggregate the trees data

Aggregate the trees dataframe by the idp column (sampled point id) to retrieve the sum of vHa and the number of trees for each sampled point. Call the aggregated dataframe pointsVol and its two columns, vHa and ntrees.

---
```
# [...] Write your code here
```
---

## 11.5.2.7 STEP 7: Join datasets

Join the output of Step 6 (the trees dataframe aggregated "by point") with the original points dataframe using the column `idp` that should be present in both the dataframes.

```
# [...] Write your code here
```

## 11.5.2.8 STEP 8: Filter data

Use boolean selection to apply the following filters to the `points` dataframe resulting from the joining in Step 7:

```
filter_nTrees          = points.ntrees .> 5 # you skip points
                                            with few trees
filter_IHaveAgeClass   = .! in.(points.cac,Ref(["AA","NR"]))
# Exclude points without age class information
filter_IHaveMainSpecies = .! ismissing.(points.esspre)
# Exclude points without a clear main species defined (e.g.
mixed species stands)
filter_overall         = filter_nTrees .&& filter_
IHaveAgeClass .&& filter_IHaveMainSpecies
# [...] Write your code here
```

## 11.5.2.9 STEP 9: Compute the age class

Run the following command to parse the age class (originally as a string indicating the five age groups) to an integer and compute the midrange of the class in years. For example, class "02" will become 7.5 years.

```
points.cac              = (parse.(Int64,points.cac) .- 1 ) .*
                          5 .+ 2.5
```

## 11.5.2.10 STEP 10: Define the model to fit

Define the following logistic model of the volumes as a function of the age with three parameters and make its vectorized form:

```
logisticModel(age,parameters) = parameters[1]/(1+exp
(-parameters[2] * (age-parameters[3]) ))
logisticModelVec(age,parameters) = # [...] Write your code
here. Broadcast the function `logisticModel` with respect to
the age, but not with respect to the `parameters` array
```

## 11.5.2.11 STEP 11: Set the initial values for the parameters to fit

Set initialParameters as a vector with 1000, 0.05, and 50, respectively. These parameters are used only as a starting point and represent the max volumes (per hectare), the growth rate of the volumes, and the ages of the trees at the point where the volumes reach half the maximal values. The actual values are those retrieved from the fit step in Step 12.

```
# [...] Write your code here
```

# CHAPTER 11  SCIENTIFIC LIBRARIES

## 11.5.2.12 STEP 12: Fit the model

Perform the fitting of the model `logisticModelVec` using the function `curve_fit(model,Xs,Ys,initial parameters)` and obtain the "fitted object" `fitobject`. From this object, extract the fitted parameters.

---
*# [...] Write your code here*

---

## 11.5.2.13 STEP 13: Compute the errors

Compute the standard error for each estimated parameter using the function `stderror(fitted object)`, and compute the confidence interval at 10% significance level with the function `confidence_interval(fitted object, significant level)`.

---
*# [...] Write your code here*

---

## 11.5.2.14 STEP 14: Plot fitted model

Plot a chart of fitted volumes (y) by stand age (x) (i.e., the function `logisticModel` with the given parameters).

---
*# [...] Write your code here*

---

### 11.5.2.15 STEP 15: Add the observations to the plot

Add to the plot a scatter chart of the actual observed VHa.

```
# [...] Write your code here
```

### 11.5.2.16 STEP 16: Differentiate the model per tree species

Look at the growth curves of individual species. Try to do the previous analysis for individual species; for example, plot the fitted curves for the five most common species.

```
# [...] Write your code here
```

## 11.5.3 Results

In Step 14 you should get a sinusoidal curve with errors spread very widely over the model. This is normal, as considering a single model for the different forest species will inevitably lead to large errors. As Step 16 shows, different tree species have very different growth models.

## 11.5.4 Possible Variations

You might consider plotting the errors of the individual tree species models, or plotting the confidence intervals of the model. You could also go further and consider environmental characteristics described in the points database, perhaps at this point using a machine learning approach (see Chapter 12).

# CHAPTER 12

# AI with Julia

The following third-party package is covered in this chapter:

| BetaML.jl | https://github.com/Sylvaticus/BetaML.jl | v0.12.1 |

This chapter gives you a brief introduction to artificial intelligence tasks with Julia. It first introduces the related concepts of artificial intelligence and machine learning, providing an overview of their scope and approaches. It then introduces BetaML.jl (https://github.com/Sylvaticus/BetaML.jl), the machine learning library you will use in this chapter, and its general organization. The chapter then presents examples of machine learning tasks, focusing first on data preprocessing, then on model training, and finally on model evaluation, interpretation, and hyperparameter tuning. It concludes with a list of other Julia ML libraries that are useful for specific tasks.

I emphasize that the data used in this chapter are synthetic and at the smallest possible level, so that they remain useful to highlight the operation described. You can find real-world examples in the exercises. Similarly, in most cases the models are shown with their default parameters or a limited subset of their options. Refer to the individual model documentation (and in particular the associated hyperparameters structure) for a complete list of the options they take.

CHAPTER 12   AI WITH JULIA

# 12.1  Machine Learning Goals and Approaches

While many authors introduce machine learning (ML) to be a subset of artificial intelligence (AI), I consider the two terms as different perspectives of the same subject. I use the term AI specifically to emphasize the high-level functions that ML algorithms can provide. But in practice, there is no form of AI that does not depend on ML. Whether we are talking about image recognition, natural language processing (e.g., chatbots or translators), or autonomous robots, they all use ML techniques. ML is characterized by problem-solving approaches that don't try to manually hard-code the process-based rules of the system under analysis. For example, they don't try to recognize the characters in an image based on predefined rules about the line patterns in the image. Instead, they "learn" these rules from the data itself. The good news is that the ML algorithms that allow this "learning" step to take place are generic and can adapt to very different problems. The number of general-purpose ML algorithms that fit a wide range of problems is surprisingly small.

Traditionally, ML has been divided into three categories.

In *supervised learning*, the ML algorithm has to predict some unknown information about the data, the *label* (often referred to as y), from some other known information, the *features* (often referred to as x). If the label is a continuous variable, the task is a *regression*; if the label is categorical, the task is a *classification*. For example, you may be interested in predicting the quality of a certain wine from its chemical and physical properties. Or the kind of objects depicted in an image, given a pixel representation of the image. To do either, you would give the ML algorithm a *training* set of data, with both the features and the labels. The algorithm would then try to discover the relationship between them, the main challenge being to find a relationship that is not specific to the training data itself but can generalize to new, unseen data. To evaluate the quality of the trained model, you would use a *test set* that was not used during model training, but for which you have the (x,y) tuple, and you would compare the y values predicted by the model with the actual values from the data.

In *unsupervised learning*, the ML algorithm still needs to discover some new information, but this time without the true values that were available in the supervised learning approach. In practice, this new information takes the form of some inherent structure in the data that is difficult to detect manually—for example, due to the high dimensionality of the data. Unsupervised algorithms can be used to separate the data into different categories (*clustering*) or to reduce the dimensionality by finding simple structures that still well represent the information embedded in the data (*dimensionality reduction algorithms*).

Finally, *reinforcement learning* is used to provide an agent with the optimal path to a given goal, which is reached after a certain sequence of steps. The goal can be, for example, winning a game of chess or guiding an autonomous robot in a physical environment. Similar to supervised learning, reinforcement learning provides feedback to the algorithm, but this feedback only concerns the very last state resulting from the set of decisions made by the algorithm. So there will be many trials, or "games," each consisting of different actions that the algorithm takes, and each action changes the current state of the world until we reach a final state, at which point the algorithm is given feedback on how good that "final" state is. The algorithm then has to work out how the individual decisions it has made relate to the final outcome.

While division of ML into these three categories remains useful for didactic purposes, modern ML development is often at the intersection of these tasks. For example, large language models such as ChatGPT and Gemini often use a mixture of all these approaches.

CHAPTER 12   AI WITH JULIA

## 12.2 The BetaML Toolkit

In this chapter, you will use BetaML.jl (https://github.com/Sylvaticus/BetaML.jl), a Julia package that provides machine learning algorithms written in the Julia programming language[1]. Its development is led by the Bureau d'Economie Théorique et Appliquée (BETA) in France, and I am one of the main contributors. I developed it and then selected it for this book for several reasons. The first reason is that BetaML provides a wide range of ML algorithms, including supervised and unsupervised learning and data processing utilities, making it suitable for building complete and diverse machine learning workflows. You will be introduced to many of these algorithms in this chapter. Secondly, BetaML uses an intuitive and consistent application programming interface (API), which makes it suitable for both novice and experienced ML practitioners working in Julia. Finally, the availability of the source code in Julia allows you to understand how the algorithms work and possibly modify them to suit your specific needs.

The following is a list of the models offered by BetaML at the time of writing:

- **Supervised models**:
  - Perceptron-like classifiers (mostly linear): PerceptronClassifier (https://sylvaticus.github.io/BetaML.jl/stable/Perceptron.html#BetaML.Perceptron.PerceptronClassifier), KernelPerceptronClassifier (https://sylvaticus.github.io/BetaML.jl/stable/Perceptron.html#BetaML.Perceptron.KernelPerceptronClassifier),

---

[1] Lobianco, A., (2021), "BetaML: The Beta Machine Learning Toolkit, a self-contained repository of Machine Learning algorithms in Julia." *Journal of Open Source Software*, 6(60), https://doi.org/10.21105/joss.02849.

CHAPTER 12  AI WITH JULIA

PegasosClassifier (https://sylvaticus.
github.io/BetaML.jl/stable/Perceptron.
html#BetaML.Perceptron.PegasosClassifier)

- Tree-like estimators: DecisionTreeEstimator
(https://sylvaticus.github.io/BetaML.
jl/stable/Trees.html#BetaML.Trees.
DecisionTreeEstimator), RandomForestEstimator
(https://sylvaticus.github.io/BetaML.
jl/stable/Trees.html#BetaML.Trees.
RandomForestEstimator)

- Neural network estimator:
NeuralNetworkEstimator (https://sylvaticus.
github.io/BetaML.jl/stable/Nn.html#BetaML.
Nn.NeuralNetworkEstimato)

- **Unsupervised models**:

    - Clustering: KMeansClusterer (https://
    sylvaticus.github.io/BetaML.jl/stable/
    Clustering.html#BetaML.Clustering.
    KMeansClusterer), KMedoidsClusterer
    (https://sylvaticus.github.io/BetaML.jl/
    stable/Clustering.html#BetaML.Clustering.
    KMedoidsClusterer), GaussianMixtureClusterer
    (https://sylvaticus.github.io/BetaML.
    jl/stable/GMM.html#BetaML.GMM.
    GaussianMixtureClusterer)

    - Dimensionality reduction: PCAEncoder (https://
    sylvaticus.github.io/BetaML.jl/stable/Utils.
    html#BetaML.Utils.PCAEncoder), AutoEncoder
    (https://sylvaticus.github.io/BetaML.jl/
    stable/Utils.html#BetaML.Utils.AutoEncoder)

- **Data and workflow processing**:
  - Scalers: MinMaxScaler (https://sylvaticus.github.io/BetaML.jl/stable/Utils.html#BetaML.Utils.MinMaxScaler), StandardScaler (https://sylvaticus.github.io/BetaML.jl/stable/Utils.html#BetaML.Utils.StandardScaler)
  - Encoders: OneHotEncoder (https://sylvaticus.github.io/BetaML.jl/stable/Utils.html#BetaML.Utils.OneHotEncoder), OrdinalEncoder (https://sylvaticus.github.io/BetaML.jl/stable/Utils.html#BetaML.Utils.OrdinalEncoder)
  - Missing imputation: SimpleImputer (https://sylvaticus.github.io/BetaML.jl/stable/Imputation.html#BetaML.Imputation.SimpleImputer), GaussianMixtureImputer (https://sylvaticus.github.io/BetaML.jl/stable/Imputation.html#BetaML.Imputation.GaussianMixtureImputer), RandomForestImputer (https://sylvaticus.github.io/BetaML.jl/stable/Imputation.html#BetaML.Imputation.RandomForestImputer), GeneralImputer (https://sylvaticus.github.io/BetaML.jl/stable/Imputation.html#BetaML.Imputation.GeneralImputer)
  - Model evaluation and analysis: ConfusionMatrix (https://sylvaticus.github.io/BetaML.jl/stable/Utils.html#BetaML.Utils.ConfusionMatrix), FeatureRanker (https://sylvaticus.github.io/BetaML.jl/stable/Utils.html#BetaML.Utils.FeatureRanker)

While using `BetaML` is practical and convenient for many tasks, for complex deep learning tasks or large-scale projects, specialized packages typically offer a wider range of tools and more computationally efficient algorithms. See the last section of this chapter for a list of alternative Julia ML packages.

## 12.2.1 API and Key Principles

All models use the same API, inspired in some ways by the Python scikit-learn library.

First, you need to load the library with `using BetaML` and then create a model object using its constructor, passing it the parameters you want:

`mod = model name(par1=a,par2=b,...)`

These parameters can be either general options common to many models, such as the level of verbosity (`verbosity`, either `NONE`, `LOW`, `STD`, `HIGH`, or `FULL`) or the random number generator (`rng`), or they can be the so-called *hyperparameters*, model-specific settings that influence the nature of the learning step. I will discuss hyperparameters later in the section "Model Evaluation, Interpretation, and Hyperparameter Tuning."

Sometimes a parameter is itself another model, in which case you would have

`mod = ModelName(par1=OtherModel(a_par_of_OtherModel=a,...),par2=b,...).`

Once you have a model, you can fit it to the data, using either the features and the labels if the model is supervised or just the features for unsupervised models:

`fit!(mod,x,[y])`

For online algorithms—models that support updating the learned parameters with new data—`fit!` can be repeated as new data arrives.

## CHAPTER 12  AI WITH JULIA

All BetaML functions work with standard Julia arrays, with records (observations) as rows and features (dimensions or categories) as columns. If you are working with DataFrames, refer to the example in Chapter 10 on how to convert DataFrames to matrices for use with BetaML.

Fitted models can be used to predict the label when given new data features:

ŷ = predict(mod,x)

As a convenience, the predictions of the last training are kept in the model object, unless explicitly opted out with cache=false in the model constructor. The predictions can simply be retrieved with predict(mod). Again, the fit! function returns ŷ instead of nothing, effectively making it behave like a fit-and-transform function. The following three expressions are therefore equivalent:

---
ŷ  = fit!(mod,xtrain,[ytrain])
ŷ1 = predict(mod)
ŷ2 = predict(mod,xtrain)
---

 **Note**

There is no formal distinction in BetaML between a data transformer, or even a model that measures the quality of another model, and an unsupervised model. They are all treated as unsupervised models that learn, given some data, how to return some useful information, regardless of whether that information is a class grouping, a particular transformation, or a quality score.

## 12.2.1.1 Managing Stochasticity

Machine learning workflows include stochastic components in several steps: in data sampling, in model initialization, and often in the model's own algorithms (and sometimes in the prediction step). All BetaML models with a stochastic component support an rng parameter, a *Random Number Generator (RNG)*, which is set by default to Random.GLOBAL_RNG. Recall from Chapter 2 that an RNG is a "machine" that streams a sequence of pseudo-random numbers depending on its seed.

If you want to "fix" only some parts of your ML workflow, you can create your model object or call individual functions with rng = FIXEDRNG, where FIXEDRNG is an instance of StableRNG(FIXEDSEED) provided by BetaML. Use it with

- MyModel(;rng=FIXEDRNG): Always produces the same sequence of results on each run of the script ("pulls" from the same rng object on different calls)

- MyModel(;rng=StableRNG(SOMEINTEGER)): Always produces the same result (new identical rng object on each call)

This option is very handy, especially during model development: a model using (...,rng=StableRNG(some_integer)) will produce stochastic results that are isolated, because the model doesn't depend on the consumption of the random stream by other parts of the workflow.

In particular, use rng=StableRNG(FIXEDSEED) or rng=copy(FIXEDRNG) with FIXEDSEED to retrieve the exact output as in the BetaML documentation, unit tests, and in this chapter's examples.

**Note**

The rng parameter has been omitted from the text for clarity, but it is present in the source files in the book's GitHub repository. If you omit it, your results will likely be different from the output shown in this chapter.

Most of the stochasticity occurs when *training* a model. However, in a few cases (e.g., decision trees with missing values), some stochasticity also occurs when *predicting* new data with a trained model. In such cases, the model doesn't constrain the random seed, so you can choose to use a fixed or variable random seed at the time of *prediction*.

Finally, if you plan to use multiple threads and want to have the same stochastic output regardless of the number of threads used, have a look at the documentation for generate_parallel_rngs.

### 12.2.1.2 Other Functions

All trained models can be reset with reset!(mod) to discard the learned information. Training information other than the parameters learned by the algorithm can be retrieved with info(mod), which returns a dictionary encoded with model-specific information.

Hyperparameters, options, and learned parameters can be obtained with the functions hyperparameters, parameters, and options, respectively. Note that they can also be used to set new values for the model, as they return a reference to the required objects.

Some models allow an inverse transformation, inverse_predict(mod,xnew). Using the parameters learned at training time (e.g., the scale factors), inverse_predict performs an inverse transformation of new data into the space of the training data (e.g., the unscaled space).

Trained models can be saved to disk with model_save("fitted_models.jld2"; mod) and loaded with mod = model_load("fitted_models.jld2","mod"). The advantage over the serialization functionality available in the Julia core is that the two functions are actually wrappers around equivalent JLD2 (https://juliaio.github.io/JLD2.jl/stable/) package functions, and should maintain compatibility across different versions of Julia.

## 12.3 Data Preprocessing

Typically, most of an ML workflow is spent preprocessing the data. See Chapter 10 for details on preprocessing related to streamlining the data so that the dataset is in a "tidy" format for analysis, where each row is a different data set (observation) and each column is a given characteristic (dimension or feature).

Even if the dataset is "tidy," you may need to preprocess the data further, depending on the ML algorithm you want to use. In this section, I will show you how to perform some specific common operations, namely encoding categorical values, scaling numerical values, imputing missing values, and two techniques for reducing the dimensionality of the data. I will also cover the partitioning of the data into the training and test sets (and eventually the validation set).

### 12.3.1 Encoding Categorical Data

Many powerful ML algorithms only work with numerical data, but your dataset may contain some categorical variables, such as color, gender, species, and so on. In such cases, you need to encode the variables into numerical values. Two simple common encoding techniques are *one-hot encoding*, which converts each possible category into a single column (dimension) with a zero/one dummy, and *ordinal encoding*, which converts the categories into integer values in a single column. Here is an example:

## CHAPTER 12 AI WITH JULIA

```
x = ["blue","red","blue","green","red"]
m1 = OneHotEncoder()
m2 = OrdinalEncoder()
x_oh  = fit!(m1,x)
x_ord = fit!(m2,x)

julia> x_oh
5×3 Matrix{Bool}:
 1  0  0
 0  1  0
 1  0  0
 0  0  1
 0  1  0
julia> x_ord
5-element Vector{Int64}:
 [1, 2, 1, 3, 2]
```

Both encoders work on a single vector. If you have several columns to encode, you can simply loop over them and use several separate models; for example:

```
x = ["blue" "apple"; "red" "apple"; "blue" "orange"; "green" "orange"]
enc_models = [OneHotEncoder() for i in axes(x,2)]
x_oh = hcat([fit!(enc_models[i],x[:,i])  for i in axes(x,2)]...)
```

276

```
julia> x_oh
4×5 Matrix{Bool}:
 1  0  0  1  0
 0  1  0  1  0
 1  0  0  0  1
 0  0  1  0  1
```

---

Calling info(model) returns some basic information:

---

```
julia> info(m1) # same for m2
Dict{String, Any} with 2 entries:
  "n_categories"   => 3
  "fitted_records" => 5
```

---

Both encoders have options to handle the case of possible categories not seen during the training of the encoder. Both also support inverse_predict to retrieve the original categories. These are stored internally in the fitted model object:

```
inverse_predict(m1,x_oh) == inverse_predict(m2,x_ord) == x # true
```

## 12.3.2 Scaling

A second common preprocessing step is to standardize the data between the different columns, so that the different features don't have completely different magnitudes.

BetaML provides a Scaler wrapper for two different models: StandardScaler (used by default), which scales each feature to unit variance and zero mean, and MinMaxScaler, which scales to a unit

## CHAPTER 12   AI WITH JULIA

hypercube. You define the model, and possibly its model-specific options, as the first option or with the `method` keyword of the `Scaler` model wrapper. Here is an example:

```
x  = [5000,1000,2000,3000]
m1 = Scaler() # eq. to Scaler(method=StandardScaler(scale=true,
              center=true))
m2 = Scaler(method=MinMaxScaler())
x_ss = fit!(m1,x)
x_mm = fit!(m2,x)
julia> x_ss
4-element Vector{Float64}:
[1.5212, -1.1832, -0.5071, 0.1690]
julia> x_mm
4-element Vector{Float64}:
[1.0, 0.0, 0.25, 0.5]
```

You can check that `x_ss` is scaled correctly:

```
using Statistics, Test
@test isapprox(mean(x_ss),0,atol=1E-15)
@test var(x_ss,corrected=false) ≈ 1
```

Both models support `inverse_predict`:

```
inverse_predict(m1,x_ss) == inverse_predict(m2,x_mm) == x # true
```

Scaler can work on whole matrices and has an optional `skip` parameter for the positional IDs of the columns to skip scaling (e.g., categorical columns, dummies, etc.):

```
x   = [[5000,1000,2000,3000] ["a", "b", "c", "d"] [5,1,2,3]
       [0.5, 0.1, 0.2, 0.3]]
m   = Scaler(skip=[2])
x_ss = fit!(m,x)
julia> x_ss
4×4 Matrix{Any}:
  1.52128    "a"    1.52128     1.52128
 -1.18322    "b"   -1.18322    -1.18322
 -0.507093   "c"   -0.507093   -0.507093
  0.169031   "d"    0.169031    0.169031
```

## 12.3.3 Missing Value Imputation

You have already seen a basic example of missing value imputation in Chapter 10. Apart from `SimpleImputer`, all BetaML imputation models work by trying to find a connection between each specific column of the data table, considered as y, and its complement columns, considered as x, using a supervised model trained on the non-missing values in y. Once the column-specific model is trained, it is used to predict the missing values in y. Because the missing values may be in different columns, this process is performed on all columns with missing values and may be iterated for several passages in the dataset. To account for the uncertainty introduced by imputation, imputation models can also be set to generate multiple imputations for the same value, similar to the R package `mice`.

The imputation models discussed next are available in `BetaML`.

## 12.3.3.1 SimpleImputer

`SimpleImputer` simply imputes the missing value within a variable with a statistic of the variable across all the records (the mean by default). It can optionally use the p-norm of the dataset (rows) to account for heterogeneity between different records. This would be appropriate in cases with large differences in the observations; for example, if the data represent the export quantities of different products per country, where some countries are very small and others are very large. Here is an example:

```
x = [2.0 missing 10; 20 40 100; ]
m1 = SimpleImputer()
m2 = SimpleImputer(norm=1)
x_full1 = fit!(m1,x)
x_full2 = fit!(m2,x)
julia> x_full1
2×3 Matrix{Float64}:
   2.0   40.0    10.0
  20.0   40.0   100.0
julia> x_full2
2×3 Matrix{Float64}:
   2.0    4.04494   10.0
  20.0   40.0      100.0
```

## 12.3.3.2 GaussianMixtureImputer

As the name implies, the `GaussianMixtureImputer` model uses a Generative (Gaussian) Mixture Model (GMM) as the underlying data-generating model, estimated by an expectation-maximization algorithm (discussed later). After fitting the mixture to the non-missing data, the missing data are computed using the Total Expectation Theorem, which weights the expected values of each mixture by the probability that the particular record belongs to that mixture. GaussianMixtureImputer shares the options with the other GMM-based models, enabling you to specify, among many other options, the number of classes to use in the mixture estimation, the family of mixtures to use (currently scalar, diagonal, or full variance Gaussians), or the initial mixtures to use, including their (initial) means and variances.

For example:

```
x = [10 2.5; missing 20.5; 0.8 18; 0.4 22.8; 12 missing; 9 3.7];
m = GaussianMixtureImputer(n_classes=2)
x_full = fit!(m,x)
julia> x_full
6×2 Matrix{Float64}:
 10.0   2.5
  0.6  20.5
  0.8  18.0
  0.4  22.8
 12.0   3.1
  9.0   3.7
```

Because the mixture initialization is deterministic and there is no other stochastic component in the model, GaussianMixtureImputer doesn't provide multiple imputations. However, it does provide metrics of the quality of the imputation based on likelihood, such as the Bayesian Information Criterion (BIC) or the Akaike Information Criterion (AIC), both of which can be obtained by querying the trained model with info:

```
julia> info(m)
Dict{String, Any} with 7 entries:
  "xndims"           => 2
  "error"            => [2.22806, 0.391854, 0.0195655,
                         1.5742e-18]
  "AIC"              => 134.569
  "fitted_records"   => 6
  "lL"               => -58.2844
  "n_imputed_values" => 2
  "BIC"              => 132.695
```

In particular, BIC and AIC trade off a higher likelihood that you can obtain when the number of classes increases with the number of estimated parameters that you will need.

### 12.3.3.3 RandomForestImputer

RandomForestImputer is probably the most accurate imputer in the toolbox, but also the slowest. It is also the only imputer that works on any kind of tabular data, including a mix of categorical and numerical columns. You can pass it either specific random forest options (which you will see in the upcoming section "Model Fitting") or imputer options, such as the number of recursive passages (recursive_passages) and multiple imputations (multiple_imputations). By default, all columns with missing values are imputed, but you can specify only a subset of them if needed (cols_to_impute).

In the following snippet, you have a matrix with missing data and a mix of numeric and categorical data, where RandomForestImputer has automatically chosen a regression or classification strategy for the different columns. Because you are using multiple imputations, the result of the `fit!` call is a vector of matrices, and the last messy line simply distinguishes between the numeric columns, where it takes the median of all imputations, and the categorical columns, where it takes the mode.

---

```julia
using Statistics
x = [1.4 2.5 "a"; missing 20.5 "b"; 0.6 18 missing; 0.7 22.8 "b"; 0.4 missing "b"; 1.6 3.7 "a"]
m = RandomForestImputer(multiple_imputations=10,recursive_passages=3)
x_full_v = fit!(m,x) # returns a vector of imputed matrices
x_full   = [nonmissingtype(eltype(identity.(x[:,c]))) <: Number ? median([v[r,c] for v in x_full_v]) : mode([v[r,c] for v in x_full_v]) for r in axes(x,1), c in axes(x,2)]
julia> x_full
6×3 Matrix{Any}:
 1.4       2.5      "a"
 0.601083  20.5     "b"
 0.6       18.0     "b"
 0.7       22.8     "b"
 0.4       20.2146  "b"
 1.6       3.7      "a"
```

---

## 12.3.3.4 GeneralImputer

GeneralImputer is similar to GaussianMixtureImputer and RandomForestImputer in that it uses a supervised model to impute the data, but it allows you to choose any model, not necessarily one from the BetaML toolkit. All that is required is that the supervised model implements (or is wrapped by) a similar API, with the model constructor, a function to fit the model, and another function to predict new x values.

## CHAPTER 12   AI WITH JULIA

Here you use `GeneralImputer` with the BetaML `NeuralNetworkEstimator`:

```
x = [1.4 2.5 1; missing 20.5 2; 0.6 18 2; 0.7 22.8 2; 0.4 missing 2; 1.6 3.7 1; ]
m = GeneralImputer(estimator=NeuralNetworkEstimator(),fit_function = BetaML.fit!, predict_function=BetaML.predict,)
x_full = fit!(m,x)
julia> x_full
6×3 Matrix{Float64}:
 1.4  2.5       1.0
 0.0  20.5      2.0
 0.6  18.0      2.0
 0.7  22.8      2.0
 0.4  0.602904  2.0
 1.6  3.7       1.0
```

If a single estimator is provided (as in this case), `GeneralImputer` will duplicate it for the columns to be imputed. Alternatively, you can provide a vector or (possibly) different imputers to, for example, impute numeric and categorical variables at the same time.

Note that the imputations in this example are not very convincing and you may get very different results (or, rarely, also an error). As you will see in the next section, neural networks are very powerful tools, but they may require more training data than other methods.

## 12.3.4 Dimensionality Reduction

BetaML provides two alternative models to reduce the dimensionality of the data:

- PCAEncoder performs Principal Component Analysis (PCA) to linearly project the data into a latent space that maximizes variance. Its main advantage is that it is relatively fast for small matrices and allows the number of dimensions in the output to be chosen depending on the maximum unexplained variance that is considered acceptable.

- AutoEncoder instead uses neural networks trained to represent the data itself, but with an inner layer used to reduce the dimensions. Its advantages are that the transformation applied can be nonlinear, and that there is explicitly an encoding part separate from the decoding part, allowing you to *encode* the data into the latent space, but then *decode* when you need to convert this latent space back into the original.

Consider this example to reduce the dimensionality of the data:

```
x = [0.12 0.31 0.29 3.21 0.21;
     0.22 0.61 0.58 6.43 0.42;
     0.51 1.47 1.46 16.12 0.99;
     0.35 0.93 0.91 10.04 0.71;
     0.44 1.21 1.18 13.54 0.85]
m_pca   = PCAEncoder(encoded_size=1)
m_ae    = AutoEncoder(encoded_size=1,epochs=800)
xl_pca  = fit!(m_pca,x)
xl_ae   = fit!(m_ae,x)
```

CHAPTER 12   AI WITH JULIA

Given the very small size of this example, you need to specify the autoencoder to run for many more epochs (i.e., the number of times you go through all the training data) than would be necessary in real-world analyses. For the `PCAEncoder` model only, you can specify `max_unexplained_var` as an alternative to `encoded_size`; for example, `0.01` means to return a projection with the minimum number of dimensions such that the explained variance remains above `0.99`. For the autoencoder, you can specify the layers to be used in `e_layers` and `d_layers` (you will see neural network layers in action in the "Neural Networks" section).

The latent space for the two encoders is very similar (net of the sign) in this case (see Figure 12-1):

```
julia> xl_pca
5×1 Matrix{Float64}:
[3.247; 6.502; 16.291; 10.154; 13.679;;]
julia> xl_ae
5×1 Matrix{Float64}:
[-4.392; -8.433; -20.562; -12.952; -17.310;;]
```

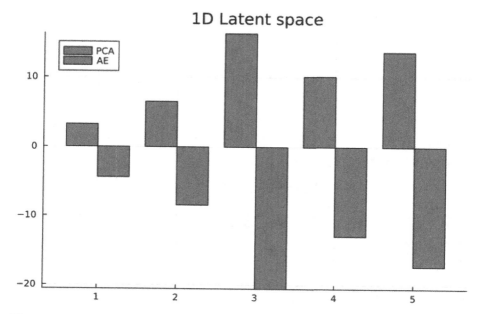

***Figure 12-1.*** *PA and PCA latent spaces*

The proportion of variance explained for the PCA and the relative mean error for the autoencoder allow you to assess the "quality" of the encoding:

```
julia> info(m_pca)
Dict{String, Any} with 5 entries:
  "explained_var_by_dim" => [0.99993, 0.999982, 1.0, 1.0, 1.0]
  "fitted_records"       => 5
  "prop_explained_var"   => 0.99993
  "retained_dims"        => 1
  "xndims"               => 5
```

```
julia> info(m_ae)
Dict{String, Any} with 11 entries:
  "par_per_epoch"   => Any[]
  "final_loss"      => 0.0156458
  "fitted_records"  => 5
  "nLayers"         => 6
  "nDLayers"        => 3
  "nepochs_ran"     => 800
  "rme"             => 0.0218386
  "nPar"            => 424
  "loss_per_epoch"  => [61.696, 61.371 … 0.016, 0.016]
  "nELayers"        => 3
  "xndims"          => 5
```

---

Finally, for the m_ae model, you can use invers_predict to project the latent space back into the original space:

```
x_reconstr = inverse_predict(m_ae,xl_ae)
julia> x_reconstr
5×5 Matrix{Float64}:
 -0.076  0.330  0.081   3.169  0.195
  0.096  0.611  0.446   6.411  0.404
  0.615  1.453  1.541  16.143  1.032
  0.289  0.925  0.854  10.037  0.638
  0.476  1.227  1.248  13.534  0.864
```

---

## 12.3.5 Data Partitioning

We conclude the data preprocessing section with data partitioning, a functionality that, unlike the models previously discussed, is implemented as a single function in BetaML: partition(data,parts;shuffle=true, dims=1,copy=true,rng).

The following example calls partition with an array of arrays (e.g., x, the version of x one-hot encoded, and y) and specifies to partition each of them into two parts of 80% and 20% of the records. The value returned is a tuple of three elements (because three arrays are passed), each made up of a tuple of two elements (because partitioning into two parts is specified). You can easily unpack this output into different variables:

```
x   = [0.1 0.1 3; 0.2 0.2 2; 0.3 0.3 1; 0.4 0.4 2]
xoh = [0.1 0.1 0 0 1; 0.2 0.2 0 1 0; 0.3 0.3 1 0 0; 0.4 0.4 0 1 0]
y   = ["a", "b", "c", "d"]
((x_train, x_test), (xoh_train, xoh_test), (y_train, y_test)) = partition([x,xoh,y], [0.8,0.2])
```

By default, the partition is row-wise, the arrays are shuffled (keeping the same order between arrays), and the data is copied. The partition results in the following arrays:

```
julia> x_train
3×3 Matrix{Float64}:
 0.1  0.1  3.0
 0.2  0.2  2.0
 0.3  0.3  1.0
julia> x_test
1×3 Matrix{Float64}:
```

```
 0.4  0.4  2.0
julia> y_test
1-element Vector{String}:
 "d"
```

## 12.4 Model Fitting. An Overview of the Main Algorithms

### 12.4.1 Perceptron-Like Classifiers

*Perceptron* is one of the simplest and oldest machine learning classifiers. Because of its computational efficiency, it is still used when the data is known to be linear. Crucially, the algorithm lacks a so-called regularization component to prevent overfitting (as found in support vector machines, another type of linear ML algorithm), and the only parameter left to limit overfitting is the number of epochs through which the algorithm is run.

In addition to the original perceptron algorithm (PerceptronClassifier), BetaML offers a version (KernelPerceptronClassifier) that uses the so-called "kernel trick" to allow nonlinearity in the relationship between x and y to be taken into account, although it is still up to the user to consider the "right" kernel to use, whereas in more modern algorithms—such as tree-based or neural networks—this is learned from the data. Finally, a third version, PegasosClassifier, uses a learning mechanism based on gradient descent (which will be discussed later in the chapter). While originally proposed as dichotomic classifiers (i.e., able to discriminate between only two classes), these models have been extended in BetaML to multiclass classification using the *one-vs.-all* approach, where each possible prediction is compared to all other classes together and a probability of belonging to each class is returned.

Consider the following example (see Figure 12-2):

```
x     = [2.2 28.4; 3.5 15.4; 2.5 31.2; 4.3 23.1; 5.7 12.8; 0.8
        18.9; 0.3 13.4]
ynum = x[:,1] .* 5 .- x[:,2] ./ 2 # linear relation
y     = [(i .< -4) ? "a" : ( (i .< 4) ? "b" : "c") for i
        in ynum]
colmap    = Dict("a" => :red, "b" => :green, "c" => :blue)
ycolvalues = [colmap[i] for i in y]
scatter(x[:,1],x[:,2],color=ycolvalues, label=nothing)
```

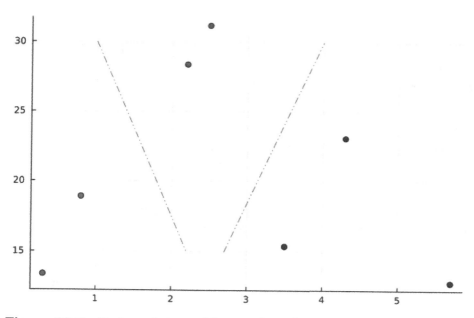

*Figure 12-2. Data points and linear classifiers*

Here you have three classes that are clearly separable with linear classifiers (i.e., in two dimensions you can draw two lines to separate the records in the three categories).

## CHAPTER 12   AI WITH JULIA

You can try the three models as follows:

---
```
m1 = PerceptronClassifier()
m2 = KernelPerceptronClassifier()
m3 = PegasosClassifier()
ŷ1_prob = fit!(m1,x,y)
ŷ2_prob = fit!(m2,x,y)
ŷ3_prob = fit!(m3,x,y)
```
---

The result for each of them is a vector, where each element (record) is a dictionary representing the estimated probability for that record to belong to that class. For example:

---
```
julia> ŷ2_prob
7-element Vector{Dict{String, Float64}}:
 Dict("c" => 0.244, "b" => 0.665, "a" => 0.090)
 Dict("c" => 0.665, "b" => 0.090, "a" => 0.244)
 Dict("c" => 0.244, "b" => 0.665, "a" => 0.090)
 Dict("c" => 0.665, "b" => 0.090, "a" => 0.244)
 Dict("c" => 0.665, "b" => 0.090, "a" => 0.244)
 Dict("c" => 0.244, "b" => 0.090, "a" => 0.665)
 Dict("c" => 0.244, "b" => 0.090, "a" => 0.665)
```
---

You can then retrieve the categories with the highest probabilities with mode(y):

ŷ3 = mode(ŷ3_prob).

I will discuss model evaluation in the next section, as this is largely independent of the specific ML algorithm, but here I anticipate that you can compute model accuracy either with the predictions in terms of probability distributions or with their modal values:

```
accuracy(y, ŷ1_prob)   # 1.0
accuracy(y, ŷ2_prob)   # 1.0
accuracy(y, ŷ3_prob)   # 0.714
accuracy(y, ŷ3)        # 0.714
```

You can see here that the first two models were able to classify the whole training set correctly, while the PegasosClassifier had some misses. You can improve the results by trying different hyperparameters of the model (e.g., in this case setting learning_rate_multiplicative=0.1) or letting BetaML find them for you (autotune=true). This is also covered in detail in the next section.

## 12.4.2 Tree-Based Models

Tree models work by fitting a decision tree. These are algorithms in which "decision rules" (the assignment of a continuous or discrete label) are made by asking a series of "questions" about the data. For example, take a look again at the last dataset:

```
julia> hcat(x,y)
7×3 Matrix{Any}:
 2.2  28.4  "b"
 3.5  15.4  "c"
 2.5  31.2  "b"
```

```
4.3    23.1    "c"
5.7    12.8    "c"
0.8    18.9    "a"
0.3    13.4    "a"
```
---

You could set a decision rule that if the first dimension is greater than or equal to `3.5`, the record should be marked `c`; otherwise, if the second dimension is greater than or equal to `28.4`, it should be marked `b`. Finally, if the second question is also false, the record should be marked `a`.

Using this decision rule, you can correctly categorize all the data. The obvious question now is how to "learn" these rules from the data itself.

The algorithm used in `BetaML` is derived from the original algorithm used in Breiman[2], where the questions are all dichotomic (i.e., there are only two possible answers, true or false) and the algorithm searches across all the dimensions, and within the dimension across all the possible specific values, for the question that divides the training set in the most homogeneous way within the two subgroups resulting from answering the question. Questions relating to categorical characteristics are in the form of "is xi equal to [something]?", whereas questions relating to continuous variables are in the form of "is xi equal to or greater than [some value]?".

For example, the question "is x1 greater than or equal to `3.5`?" splits the records into two groups, one group consisting of only three records in which y is `c`, and the other group consisting of half the records in which y is `a` and the other half in which y is `b`.

You can see that decision trees are very interpretable, but they can overfit the data; for example, an unseen record in which the first feature is `3.499` would be categorized very differently from another record in which the first feature is `3.500`. To overcome this problem of overfitting, a random

---

[2] Breiman, L., (2001), "Random Forests." *Machine Learning, 45*(5–32), `https://doi.org/10.1023/A:1010933404324`.

forest is an aggregation of the results of many different decision trees, each of which has been trained on a random subset of records and a random subset of dimensions. It turns out that random forests are very effective in classification or regression, at the cost of much lower interpretability and higher computational cost.

In BetaML you can fit a decision tree or a random forest to any type of data in the features, ordered, categorical, or even where missing values are present:

```
m1 = DecisionTreeEstimator()
m2 = RandomForestEstimator() # n_trees = 30 by default
y1hat = fit!(m1,x,y)
y2hat = fit!(m2,x,y)
m1err = accuracy(y,y1hat) # 1.0
m2err = accuracy(y,y2hat) # 1.0
```

The result is probabilistic. While the accuracy is the same in this simple example, expect better accuracy, especially for the test set, in random forests.

The use in regression is the same: the prediction returned by the algorithm is the average of the records in the group obtained by answering all the questions.

## 12.4.3 Neural Networks

Neural networks are supervised models (though they are also used for unsupervised tasks, such as autoencoders and missing-value imputers). Although the concept of neural networks was inspired by and is somewhat analogous to the neural network of the human brain, artificial neural networks are much less complex. Neural networks are really just simple transformations of data that flow from the input through various layers to an output. That's their beauty! Complex capabilities emerge from very simple units when they are combined; for example, the capability to recognize not "just" objects in an image, but also abstract concepts such as people's emotions or the message embedded in a textual prompt. Neural networks have become very popular because this "assembly" of simple computational units is highly parallelizable, allowing the use of specialized hardware (graphics processing units [GPUs], tensor processing units [TPUs], etc.), which in turn allows the efficient processing of huge data sets.

I'll first briefly describe the structure of deep neural networks and then show how to train a neural network from the data. In particular, I will discuss only the basic feed-forward networks. *Convolutional layers*, *recurrent neural networks*, and *transformers* are all different implementations (architectures) based on the same ideas discussed in this section; further discussion of them is beyond the scope of this book.

### 12.4.3.1 Deep Neural Network Structure

In *deep forward neural networks*, neural network units are arranged in *layers*, from the *input layer*, where each unit holds the input coordinate, through various *hidden layer* transformations (hence the term "deep"), to the actual *output* of the model, as depicted in Figure 12-3.

CHAPTER 12  AI WITH JULIA

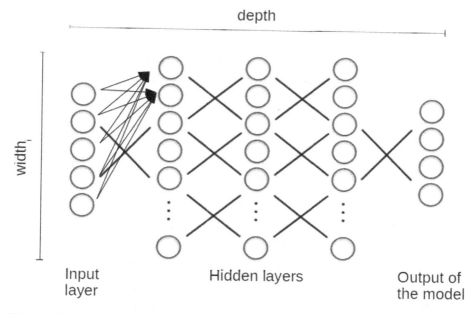

*Figure 12-3. Deep neural network schema*

Drilling down further, Figure 12-4 depicts a single *dense* neuron (in the sense that it is connected to *all* the neurons of the previous layer, or to the input layer).

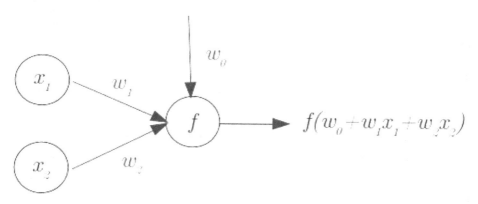

*Figure 12-4. Single-neuron schema*

The variables in Figure 12-4 are as follows:

- $x$ is a two-dimensional input, where $x_1$ and $x_2$ are the two dimensions of the input data (they could equivalently be the outputs of two previous layers of neurons).

- $w_1$ and $w_2$ are the *weights* applied to $x$ plus a constant term ($w_0$). These are the parameters that the algorithm wants to learn from the data.

- $f$ is an *activation function* (often nonlinear) applied to $w_0 + x_1 w_1 + x_2 w_2$ to define the output of the neuron.

The output of the neuron can be the final output of the neural network, or it can be the input of the next layer. For one-dimensional regression models, the last layer of the model is therefore a single neuron layer with an activation function suitable for the labels (e.g., for nonnegative values, it must be a function that maps to nonnegative values). For classification models whose data has been encoded using one-hot encoding, the last layer will have $d$ neurons, where $d$ is the number of possible categories to predict.

In `BetaML`, you build a neural network model by providing an array of the layers that make up the model and a loss function (explained in the next section). If you don't provide them, `BetaML` will try to automatically build a medium-sized network compatible with the labels and features provided during training.

```
nn1 = NeuralNetworkEstimator() # Default NN model, usable for
      both regression and classification
nn2 = NeuralNetworkEstimator(layers=[DenseLayer(6,15,f=relu),
      DenseLayer(15,1,f=relu)],loss=squared_cost) # regression
      in R+ from a 6 dims data
nn3 = NeuralNetworkEstimator(layers=[DenseLayer(6,10),Dense
      Layer(10,3),VectorFunctionLayer(3)],loss=squared_cost)
      # classification in R+ from a 6 dims data
```

CHAPTER 12   AI WITH JULIA

If you choose to specify the layers manually, the following layers are available:

- `DenseLayer`: The classic feed-forward layer with a user-defined activation function.

- `DenseNoBiasLayer`: The classic layer without the bias parameter.

- `VectorFunctionLayer`: A layer whose activation function runs over the ensemble of its neurons rather than on each neuron individually. It doesn't have learnable weights on the input, but optional learnable weights as parameters of the activation function. As in the preceding example, it is often used with the `softmax` activation function to rescale the output to a probability distribution for classification.

- `ScalarFunctionLayer`: A layer whose activation function runs over each node individually, like a classic `DenseLayer`, but with no learnable weights on the input and optional learnable weights as parameters of the activation function.

- `ReplicatorLayer`: An alias for a `ScalarFunctionLayer` with no learnable parameters and identity as an activation function.

- `ReshaperLayer`: Used to reshape the output of a layer (or the input data) into the shape needed for the next layer.

- `GroupedLayer`: A layer used to group multiple layers into a single layer. This allows for multibranch networks in which one layer connects several separate chains.

These layers have good CPU performance.

BetaML also provides the following two layers, which are currently experimental because they require performance optimizations:

- `PoolingLayer`: While the `ScalarFunctionLayer` applies the activation function scalarly to each individual input neuron, and the `VectorFunctionLayer` does it as an individual array, the `PoolingLayer` applies a function to the set of neurons defined in a sliding kernel. It is weightless.
- `ConvLayer`: A generic N+1 (channels) dimensional convolutional layer.

Each layer can use a default activation function, one of the functions provided by the `Utils` module (`relu`, `tanh`, `softmax`, etc.) or one of your own.

## 12.4.3.2 Neural Network Training

You have already seen in Chapter 9 how easy it is to implement a layer and its `forward` function in Julia. Implementing the algorithm that learns the weights is a slightly more complex task. This section will give you some insight about this, but it will not present the code (you can check the code in `BetaML`).

The basic idea of neural networks is to compute the *gradient*, the set of first derivatives, of the loss function with respect to the parameters that the algorithm needs to learn: the weights in the neuron computation or, for the layers that support them, the parameters of the activation functions. The loss function is simply a function (e.g. the Euclidean distance) that calculates how "far" the predictions made by the model are from the true labels. The gradient therefore represents the information about how each parameter contributes to the error the neural network makes in its prediction.

If the gradient is negative for a particular parameter, this means that if you *increase* that parameter slightly, you will get a lower error. Conversely, if it is positive, it means that if you *reduce* the parameter slightly, you will find a lower error.

An algorithm based on gradient descent can be used to iteratively search for the minimum error by moving the parameter *against* the gradient by a certain amount. The most basic algorithm is then $w_i^t = w_i^{t-1} - \frac{d\epsilon^{t-1}}{dw_i^{t-1} * \lambda}$, where $w_i^{t-1}$ is the value of the parameter before it is updated, $\frac{d\epsilon^{t-1}}{dw_i^{t-1}}$ is the total derivative of the loss with respect to the parameter, and $\lambda$ is the step you are willing to take against the gradient, also known as the *learning rate*. A "good" learning rate is essential: too small a learning rate would make learning slow, with the risk of getting trapped in a local minimum instead of a global one. Conversely, if the learning rate is too high, there is a risk that the algorithm will diverge (remember that derivatives are variations at the *margin*).

BetaML provides two optimization algorithms: the Stochastic Gradient Descent (SGD), which uses the previous algorithm, and the ADAM, which uses a slightly more complex but more efficient algorithm. Both by default use learning or decay rates that gradually decrease during training and can be specified as options of the neural network model; for example:

```
nn = NeuralNetworkEstimator(opt_alg=SGD(η = t -> 1/(1+t), λ=2))
```

The only remaining question is how to calculate the gradient. This can be calculated for each parameter of the network using the chain rule. The *backpropagation* algorithm is simply an efficient way of calculating it, starting from the last layer and working backward toward the head of the chain, so that the derivatives in later layers can be cached for calculating the derivatives in earlier layers using the chain rule.

To compute the total derivatives of the loss with respect to a given weight, the algorithm needs the partial derivatives of the activation function of each layer and of the loss function. In BetaML these can be provided manually in the df and dloss parameters of the layers and model, respectively. For some known functions, these can be provided automatically by matching the function with its known derivative. If the derivatives are not provided and f or loss is not in the list of known functions, the partial derivative is calculated using automatic differentiation (AD).

An important aspect to consider is that the calculation of the gradient depends on the data. If you process different data, you will get different derivatives. You can then move between two extremes: at one extreme, you compute the gradient as an average of those computed on all data points, and you apply the optimization algorithm to this average. At the other extreme, you sample randomly, record by record, and move the parameter on each record. The compromise is to partition the data into a set of *batches*, compute the average gradient of the batch, and then update the parameter using the optimization algorithm applied to that average gradient.

The "one record at a time" approach is the slowest but is also very sensitive to the presence of outliers. The "take the average of all data" approach is faster in running a given epoch, but it takes longer to converge (i.e., it requires more epochs). It also requires more memory because you have to store the gradients for all the records. So the batch approach is a good compromise. You can choose the batch size using the batch_size option:

```
nn = NeuralNetworkEstimator(batch_size=32)
```

Finally, note that neural networks have no mechanism to avoid overfitting. With large enough networks (in terms of width or depth) and enough training, you run the risk of overfitting the network to the training data and losing generality. You can control overfitting of neural networks by using the epochs parameters or by reducing the size of the network:

```
nn = NeuralNetworkEstimator(epochs=100)
```

### 12.4.3.3 Neural Network Example

You are now ready to run a complete example neural network. Compared to linear or tree-based models, neural networks require larger datasets to train.

You can then create the following synthetic dataset with a label expressed in both numerical and categorical format, with the latter being hot-coded:

```
(N,D)  = (1000,6)
x      = rand(N,D)
y      = abs.([10*r[1]-r[2]+0.1*r[3]*r[1] + sqrt(r[6]*10) for r
         in eachrow(x) ])
ysort = sort(y);
ycat  = [(i < ysort[Int(round(N/3))]) ?  "c" :  ( (i < ysort
        [Int(round(2*N/3))]) ? "a" : "b")  for i in y]
ohm   = OneHotEncoder();
yoh   = fit!(ohm,ycat)
```

## CHAPTER 12    AI WITH JULIA

You can also reuse the previously defined nn1, nn2, and nn3 models. However, if you use nn1 for one task (regression or classification), you modify and adapt this model for that specific task. Therefore, create two separate models so that you can use one model for regression and the other for classification:

---

```
nn1a = NeuralNetworkEstimator()
nn1b = NeuralNetworkEstimator()
```
---

Finally, fit the model and compute some basic metrics (more on these in the next section):

---

```
y1ahat = fit!(nn1a,x,y)
y1bhat = fit!(nn1b,x,yoh)
y2hat  = fit!(nn2,x,y)
y3hat  = fit!(nn3,x,yoh)

m1_rme   = relative_mean_error(y,y1ahat)                # 0.020
m1_accerr = accuracy(ycat,inverse_predict(ohm,y1bhat))  # 0.558
m2_rme   = relative_mean_error(y,y2hat)                 # 0.018
m3_accerr = accuracy(ycat,inverse_predict(ohm,y3hat))   # 0.975
```
---

Your results over a single run may vary due to the stochastic nature of the models, but the general message is that neural networks can be very powerful, but often require some practice to express their potential compared to tree-based models that require very little tuning

## 12.4.4 Clustering

*Clustering* refers to the partitioning of the dataset into classes (groups). It is an unsupervised task, as the algorithm doesn't have access to examples of "real" classes, but instead has to find a partitioning on its own by finding an inner structure in the data. This structure is often simply a distance metric, where the data is clustered according to its distance to some (multidimensional) points. For example, in Figure 12-5, you can intuitively distinguish three groups centered on (-5,-2), (0,6), and (2.5,0).

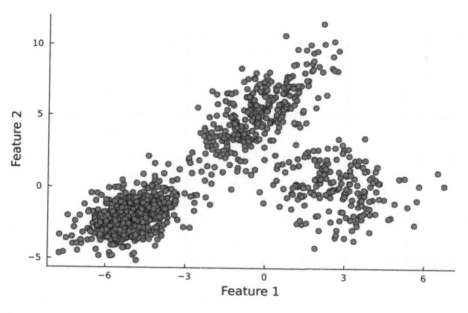

*Figure 12-5.* *Unclassified data points*

BetaML provides three clustering algorithms for these distance-based *clusterers*: KMeansClusterer, KMedoidsClusterer, and GaussianMixtureClusterer. The first two are so-called *hard clustering* algorithms, as they return a single assignment per record, while the last is a *soft clustering* algorithm, as it returns for each record a probability distribution of the record being assigned to each possible class.

## CHAPTER 12  AI WITH JULIA

In KMeansClusterer, the centers of the clusters, also known as *representatives*, can be arbitrary points, while in KMedoidsClusterer, they must be found within the records in the data. In both cases, the algorithm iterates between assigning the records to the closest representative of each class (using the provided distance metric) and calculating the representative points as the means (again using the provided metric) of the records assigned to that class.

GaussianMixtureClusterer assumes a mixture model as the underlying generative model of the data and tries to estimate it using an expectation-maximization (E-M) algorithm. In a mixture model, the observed data is assumed to be the realization of two separate random variables: first, a particular distribution (typically Gaussian) is chosen with probability $p$, and then the data is the outcome of that particular random variable. The E-M algorithm is again iterative: first, it assumes that the parameters of the different distributions are known and it maximizes the likelihood of the data to find the probability of each record belonging to a particular class, and then it takes this as a given to maximize the likelihood again, but this time in terms of the parameters of the different distributions.

For all models, you need to specify the number of classes to partition the data. In KMeansClusterer and KMedoidsClusterer, you can also specify the distance metric (default: Euclidean), while in GaussianMixtureClusterer, you can specify the type of mixtures to use (default: multivariate Gaussian with diagonal covariance matrix).

You can generate the synthetic data used to draw Figure 12-5 above with:

```
import Distributions: MixtureModel, MvNormal
# Mixture of 3 2D Normals...
data_gen_model = MixtureModel(MvNormal[
    MvNormal([2.5,0.0],[2 -0.8; -0.8 3]),
    MvNormal([-5,-2],[0.8 0.4; 0.4 1.2]),
    MvNormal([-0.5,5.2],[2 2.5; 2.5 5])], [0.2, 0.5, 0.3])
```

```
data = rand(data_gen_model,1000)'
scatter(data[:,1], data[:,2], legend=nothing, xlabel="Feature 1",
ylabel="Feature 2")
```

You can now cluster the data using the three clustering algorithms:

```
kmeans_mod    = KMeansClusterer(n_classes=3,initialisation_
                strategy="grid")
kmedoids_mod  = KMedoidsClusterer(n_classes=3,initialisation_
                strategy="grid")
gmm_mod       = GaussianMixtureClusterer(n_classes=3,mixtures=
                FullGaussian)
classes = fit!(kmeans_mod,data)
colmap  = ["red", "green","blue"]
colors  = [colmap[c] for c in classes]
scatter(data[:,1], data[:,2],color=colors, legend=nothing,
title="KMeans assignments")

classes = fit!(kmedoids_mod,data)
colors  = [colmap[c] for c in classes]
scatter(data[:,1], data[:,2],color=colors, legend=nothing,
title="KMedoids assignments")

probs   = fit!(gmm_mod,data)
classes = BetaML.mode(probs) # returns the single class with
          the higher probability
colors  = [RGB(r...) for r in eachrow(probs)];
scatter(data[:,1], data[:,2],color=colors, legend=nothing,
title="GMM assignments")
```

This results in the charts shown in Figure 12-6.

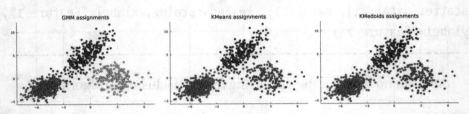

***Figure 12-6.*** *Classified points (from left: GMM, KMeans, and KMedoids)*

You can see that the assignments of KMeans and KMedoids are almost identical, while the assignments of GMM introduce some nuances in the color of the records at the border between different classes.

## 12.5 Model Evaluation, Interpretation, and Hyperparameter Tuning

In this section you will learn some metrics that allow you to judge the predictions of a given ML model. Remember that you should always make your final judgement on a supervised ML model using the test set, a subset of the data that you have reserved and haven't used for training or for hyperparameter tuning.

### 12.5.1 Regression Models

Consider the following model true labels and the predictions of an ML model:

```
y = [1.3, 16.8, 3.5]
ŷ = [0.9, 17.0, 3.2]
```

You can use norm-based distances between true values and predictions: l1_distance(y,ŷ) (a.k.a. *Manhattan distance*), l2_distance(y,ŷ) (a.k.a. *Euclidean distance*), or the cosine_distance(y,ŷ).

These measures are not scaled by the number of data records or by the mean values. Furthermore, these measures are sensitive to the scale of the data.

To scale for the number of records, use the mean squared error (MSE), mse(y,ŷ), a version of the (squared) Euclidean distance where the deviation is divided by the number of records.

To also consider the error relative to the size of the true mean values, which I find easier to communicate, look at relative_mean_error(y,ŷ), which by default uses the $\ell^1$ norm of the deviations.

Finally, perhaps an underused metric is the Sobol index sobol_index(y,ŷ), which gives the percentage of variance in the true data that is explained by the model. This metric has the advantage of being scale invariant:

```
yb  = 10 .* (y .- mean(y))
ŷb  = 10 .* (ŷ .- mean(y))
sobol_index(y, ŷ) ≈ sobol_index(yb, ŷb) # true
```

Consider now a second model whose output is ŷ2 = [1.2, 18.9, 3.6], so that you have:

```
y  = [1.3, 16.8, 3.5]
ŷ  = [0.9, 17.0, 3.2]
ŷ2 = [1.2, 18.9, 3.6]
```

Which of the two models is better? Well, it depends on the application and the nature of the data. The second model gives a larger absolute deviation, but for two out of three records, the relative error is smaller.

If you are interested in the absolute deviations, use `l1_distance`, `l2_distance`, `mse`, `relative_mean_error`, or `sobol_index`. If you are interested in models with low relative errors, use `cosine_distance` or `relative_mean_error` with the parameter `normrec=true`:

```
relative_mean_error(y, ŷ)                  # 0.042
relative_mean_error(y, ŷ2)                 # 0.106
relative_mean_error(y, ŷ; normrec=true)    # 0.135
relative_mean_error(y, ŷ2; normrec=true)   # 0.077
```

You have a similar problem when individual $y_i$ and $\hat{y}_i$ are vectors—that is, with models that estimate several variables at once. If you are interested in relative errors within the different variables, use `relative_mean_error` with the parameter `normdim=true`:

```
y  = [1.3 20.1; 4.8 18.7; 3.5 23.2]
ŷ  = [0.9 20.8; 4.0 17.8; 3.2 22.8]
ŷ2 = [1.2 23.1; 4.9 16.8; 3.6 21.8]
relative_mean_error(y, ŷ)                  # 0.049
relative_mean_error(y, ŷ2)                 # 0.092
relative_mean_error(y, ŷ; normdim=true)    # 0.094
relative_mean_error(y, ŷ2; normdim=true)   # 0.066
```

In this example, the model that produces ŷ2 makes a larger absolute error than those that produce ŷ, but it does so on the second feature, where the values are larger. In relative terms, the first model performs worse.

## 12.5.2 Classification Models

Before considering classification metrics, you should know that classification predictions can take different forms. They can be vectors of a nonnumerical nature, encoded as one-hot matrices or as integers. If the model is probabilistic, the output of each individual prediction can be either a dictionary of the probability for each label or an array describing a probability distribution.

With that in mind, take a look at the following data, where you have y and ŷ in all possible forms:

```
y = ["green", "red", "red", "green", "blue", "green"]
ohm = OneHotEncoder()
oem = OrdinalEncoder()
yoh = fit!(ohm,y) # One-hot encoded
yoe = fit!(oem,y) # Ordinal encoded (1,2,2,1,....)
levels = parameters(ohm).categories
ŷprob = [0.8 0.2 0.0
         0.4 0.3 0.3
         0.2 0.5 0.3
         0.6 0.1 0.3
         0.1 0.1 0.8
         0.3 0.4 0.2]
ŷdict = [ Dict([levels[i] => r[i] for i in
axes(ŷprob,2)])       for r in eachrow(ŷprob)]
ŷint = BetaML.mode(ŷprob)
ŷ    = BetaML.mode(ŷdict)
```

You can then calculate the accuracy in a number of ways:

```
accuracy(y, ŷ)        # 0.667
accuracy(y, ŷdict)    # 0.667
accuracy(yoe, ŷprob)  # 0.667
accuracy(yoe, ŷint)   # 0.667
```

## CHAPTER 12   AI WITH JULIA

For probabilistic models, you could also measure the average deviation of the probability distributions using cross-entropy or Kullback-Leibler divergence, considering the one-hot encoding as a degenerate probability distribution with all the probability mass concentrated in one category:

```
avg_crossentropy = sum(crossentropy(yoh[r,:], ŷprob[r,:]) for r
in axes(y,1)) / size(y,1)
avg_kl_div = sum(kl_divergence(yoh[r,:], ŷprob[r,:]) for r in
axes(y,1)) / size(y,1)
```

While the accuracy of a categorical prediction is sometimes sufficient, you often want to better investigate where your model has performed well and for which categories it has instead made more misclassifications. A detailed report is provided by the `ConfusionMatrix` model, which works on the same combinations as `Accuracy` (y vs. ŷ, y vs. ŷdict, yoe vs. ŷprob, or yoe vs. ŷint):

```
cm = ConfusionMatrix()
fit!(cm,y, ŷ)
print(cm)
```

The resulting report is as follows:

```
A ConfusionMatrix BetaMLModel (fitted)
```

```
*** CONFUSION MATRIX ***
Scores actual (rows) vs predicted (columns):
4×4 Matrix{Any}:
```

```
 "Labels"    "green"   "red"   "blue"
 "green"     2         1       0
 "red"       1         1       0
 "blue"      0         0       1
```
Normalised scores actual (rows) vs predicted (columns):

```
4×4 Matrix{Any}:
 "Labels"    "green"    "red"      "blue"
 "green"     0.666667   0.333333   0.0
 "red"       0.5        0.5        0.0
 "blue"      0.0        0.0        1.0
```

\*\*\* CONFUSION REPORT \*\*\*
- Accuracy:               0.6666666666666666
- Misclassification rate: 0.33333333333333337
- Number of classes:       3

```
  N Class   precision   recall   specificity   f1score   actual_
count   predicted_count
                        TPR      TNR                     support

  1 green   0.667       0.667    0.667         0.667     3
    3
  2 red     0.500       0.500    0.750         0.500     2
    2
  3 blue    1.000       1.000    1.000         1.000     1
    1
- Simple   avg. 0.722   0.722    0.806         0.722
- Weighted avg. 0.667   0.667    0.750         0.667
```

---

Output of `info(cm)`:
[output suppressed for space reasons]

---

From the report, you can see that this fictive model sometimes makes errors between the "green" and "red" categories. You can then analyze the situation and take corrective action; for example, if the data is unbalanced, you can rebalance the dataset.

If you prefer, you can plot the confusion matrix, the result of which is shown in Figure 12-7:

```
res = info(cm);
heatmap(string.(res["categories"]),string.(res["categories"]),
res["normalised_scores"],seriescolor=cgrad([:white,:blue]),
xlabel="Predicted",ylabel="Actual", title="Confusion Matrix
(normalised scores)")
```

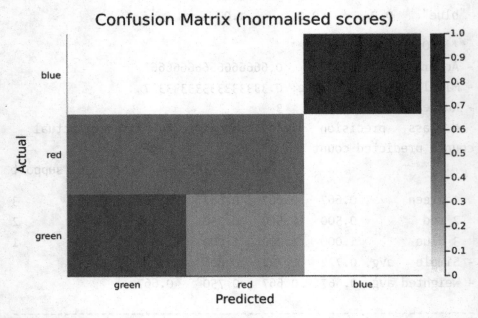

*Figure 12-7. Confusion matrix graphical output*

## 12.5.3 Clustering Models

As an unsupervised model, clustering faces the challenge that the "true" classes of the records are unknown. However, there are ways to assess the quality of a cluster, and in particular to help choose the appropriate number of classes.

One way is to compute the *silhouette values*, a measure of how similar an object is to those in its assigned class compared to those in the other classes, ranging from -1 (bad assignment) to +1 (perfect match with its own class). The silhouette values can then be plotted and manually inspected, or simply averaged to compute an average silhouette score of the entire clustering output.

For GaussianMixtureClustering, an alternative approach is to relate the estimated likelihood of the fit to the number of parameters the fit had to estimate. The idea is that the higher the likelihood and the lower the number of parameters, the better the clustering. A metric that uses this approach is the *Bayesian Information Criterion (BIC)*, where a lower value is better.

For purposes of the following example, reuse the data used in the clustering example before, keeping in mind that this data is generated from three different classes. The objective is to see how these two clustering metrics change when you try to fit the data into clusters of different sizes. In this example, you will try from two to five different cluster classes using the GaussianMixtureClusterer:

```
classes_to_test = 2:5
sil_by_class = fill(-1.0,length(classes_to_test))
BIC_by_class = fill(Inf,length(classes_to_test))
pd = pairwise(data,distance=l2_distance) # we compute the
                                           pairwise distances
```

To calculate the silhouette values, you need the pairwise distances between the data points. Since these don't change, you calculate them only once. You then loop over all possible classes and calculate the metrics:

## CHAPTER 12  AI WITH JULIA

```
for (i,cl) in enumerate(classes_to_test)
    m = GaussianMixtureClusterer(n_classes=cl,mixtures=Full
    Gaussian)
    ŷ = fit!(m,data) |> mode
    s = mean(silhouette(pd, ŷ))
    bic = info(m)["BIC"]
    sil_by_class[i] = s
    BIC_by_class[i] = bic
end
```

This results in the chart shown in Figure 12-8, where you can see that both metrics correctly suggest two as the optimal number of classes.

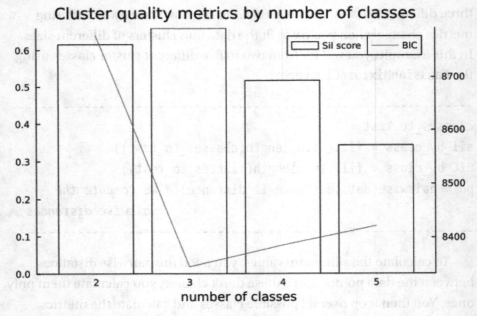

***Figure 12-8.*** *Silhouette score and BIC by number of classes considered*

## 12.5.4 Hyperparameters Evaluation

As you have seen, the quality of many machine learning algorithms depends on their hyperparameters. These are additional parameters that, unlike the normal "parameters" (which are learned from the training set), are chosen before training and remain fixed during training, but still affect the performance of the algorithm. Examples of hyperparameters include the number of neurons in a neural network layer, the number of individual decision trees in a random forest algorithm, or, for many algorithms (including perceptrons), how *long* the learning step should continue (this can take different forms depending on the algorithm).

Hyperparameters play a fundamental role in the trade-off between *specialization* and *generalization*. In fact, the goal of training is not to learn a relationship between the given x and the given y, but to learn from the given data the *general* relationship between x and y for the entire population from which x and y were sampled. And the hyperparameters should be "set" to values that maximize this objective, not to value that minimize errors in the data used to train the algorithm. If you use too many neurons, if you train too much, you would learn the *specific* relationship between x and y in the training data. However, this reflects the specific data provided, not the general population. In statistical terms, you would *overfit* your model or, in other words, generate a large *variance* in the model's trained parameter, which would be too dependent on the specific training data (i.e., different training data would lead to very different learned parameters). Conversely, if the model is too simple or receives too little training, it will have too much *bias* and will not learn the relationship sufficiently. Techniques that allow an algorithm to generalize better at the expense of better performance on the training data are called "regularization" techniques.

How can you choose the hyperparameters that minimize the bias-variance trade-off? In general, no assumptions are made about the data other than that they all come from the same population.

Aside from relying on the literature or previous experiments, the simplest approach to finding good hyperparameters is to use the data itself to "evaluate" the generality of the model by randomly partitioning the dataset into three subsets:

- **Training set**: Used to actually "train" the algorithm to learn the relationship between the given x and the given y, conditional on a given set of hyperparameters; in other words, this set is used to find the parameters that minimize the error made by the algorithm.

- **Validation set**: Used to evaluate the results of your trained algorithm on data that has not been used by the algorithm to train the parameters (out of sample); in other words, this set is used to find the hyperparameters that allow the best generalization.

- **Test set**: Used to evaluate the overall performance of the algorithm when used with the "best" hyperparameter (you can't use the validation set for this, as the hyperparameters are "fitted" based on it).

| training | validation | test |
|---|---|---|

In practice, you have several ways to search for the "best hyperparameters": grid search over the hyperparameter space, random search, gradient-based methods, and so on. In all these cases, you run the algorithm on the training set and evaluate it on the validation set until you find the "best" set of hyperparameters. At this point, with the "best" hyperparameters, you train the algorithm one last time on the training set and evaluate it on the test set.

## 12.5.5 K-fold Cross-Validation

While a single training/evaluation step for each possible hyperparameter set may sometimes be sufficient, many ML algorithms have a stochastic nature, either in the model initialization or in the training algorithm. A better approach is to have multiple instances of training and evaluation. A common technique to train and validate an ML model multiple times is *K-fold cross-validation*, where separate training and validation sets are sampled multiple times from a common training and validation set. Figure 12-9 depicts an example of 5-fold cross-validation.

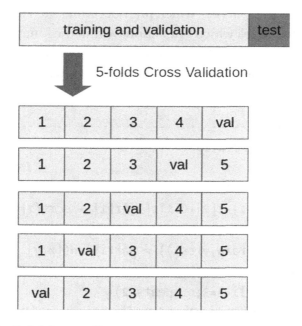

*Figure 12-9. 5-fold sampling*

First, the entire data set is randomly partitioned into a train/validation set and a test set. Then, for each possible hyperparameter set, you randomly divide the train/validation test into K subsets. You use K – 1 of the subsets for training and use the remaining subset for computing the out-of-sample score of the model. You do this (keeping the same hyperparameters and the same partition) for all the different K subsets and you compute a statistic (e.g., the average) of the performances of the model with that given hyperparameter. You then compare the statistic and select the "best" hyperparameters, perform a final training on the train/validation set and make the evaluation on the test set.

BetaML provides the function `cross_validation(f,data,sampler)`, which performs cross-validation according to the rules defined by the `sampler` object passed to it, by calling the function `f` and collecting its output. This function is very flexible because it leaves it to you to define what to do with each split (i.e., which preprocessing, such as scaling or encoding, which model, which evaluation) within the function `f`, which (conveniently) can take the form of a do block.

Consider the following example, reusing the (x,y) dataset you used in the "Neural Networks" section:

```
# Syntactic data generation
(N,D) = (1000,6)
x     = rand(N,D)
y     = abs.([10*r[1]-r[2]+0.1*r[3]*r[1] + sqrt(r[6]*10) for r
in eachrow(x) ])
((xtrain,xtest),(ytrain,ytest)) = partition([x,y],[0.8,0.2])
# Sampler object
sampler = KFold(nsplits=3;nrepeats=2)
# Cross-validation implementation for the specific
hyperparameter n_trees=30
(μ,σ) = cross_validation([xtrain,ytrain],sampler) do train_
data,val_data, rng
    (xtrain,ytrain) = train_data; (xval,yval) = val_data
    model           = RandomForestEstimator(n_trees=30)
    fit!(model,xtrain,ytrain)
```

```
    ŷval           = predict(model,xval)
    ϵ              = relative_mean_error(yval, ŷval)
    return ϵ
end # (0.0719, 0.00865)
```
---

Here you first define the sampler as an object that splits the training set into three random subsets and repeats the random split twice. Then you call `cross_validation` with this `sampler` object and with the dataset in the [x,y] array. If you were evaluating an unsupervised model—for example, a cluster model with one of the metrics previously described—you would have passed only [x]. The data you passed to the `cross_validation` function is then split according to the rules of the `sampler` object, and the data from each split is passed to the `f` function (here in the form of a do block) as separate `train_data` and `val_data`. Note that each of these is a tuple of (x,y), because originally you also passed a tuple of (x,y) to `cross_validation`. Using this split-specific data, the do block initializes the desired model, fits it to the training data, predicts based on the validation data, and computes the appropriate metric. The do block is run six times in this case (three splits by two repetitions), each time producing a possibly different error. This is collected and the mean and standard deviation are returned. You may get different results to mine, but because of the law of large numbers, the average will be much more stable than the individual model evaluations you ran before.

## 12.5.6 Autotune

Now you have a good idea of the goodness of a particular ML model with a particular set of hyperparameters. However, you still need to run cross-validation in a loop over the hyperparameters you want to test.

To make life easier, most `BetaML` models accept an `autotune` parameter that does this for you, based on certain hyperparameter ranges/sets and some search heuristics.

## CHAPTER 12 AI WITH JULIA

The search algorithm can be specified during model construction using the `tuning_method` parameter, which is set by default to `SuccessiveHalvingSearch`, an object which itself has several configurable options within those `hpranges` and `res_shares`. The first option is a dictionary of the different hyperparameters and their values to be evaluated, where each type of model has its own defaults. The second option relates to the resources you allocate to the algorithm to search for the optimal hyperparameters. In `SuccessiveHalvingSearch`, you first try all possible hyperparameters with very few resources (records), then you keep only the best-performing combinations and try them again with more records, and so on until you find the "best" combination. The `res_shares` control how many iterations to use and the share of records to use in each iteration. Here's an example that tries different hyperparameters:

```
tuning_method = SuccessiveHalvingSearch(
                hpranges     = Dict("max_depth" =>[5,10,
                nothing], "min_gain"=>[0.0, 0.1, 0.5],
                "min_records"=>[2,3,5],"max_features"=>
                [nothing,5,10,30], "n_trees"=>[10,20,30]),
                loss         = l2loss_by_cv,
                res_shares   = [0.05, 0.2, 0.3],
                multithreads = true
                )
m = RandomForestEstimator(autotune=true, tunemethod=tuning_method)
```

Note that each "trial" uses a loss function, which in turn uses the cross-validation function you saw earlier. So the number of times the model is actually fitted to the data is very high. The "autotune" that happens during the first `fit!` call can then take a considerable amount of time, even for small models:

```
ŷtrain = fit!(m,xtrain,ytrain)
```

This should print something like:

```
Starting hyper-parameters autotuning (this could take a
while..)
(e 1 / 3) N data / n candidates / n candidates to retain :
40.0    324  47
(e 2 / 3) N data / n candidates / n candidates to retain :
160.0   47   7
(e 3 / 3) N data / n candidates / n candidates to retain :
240.0   7    1
```

This output tells you that the algorithm had three iterations. In the first one, it processed 40 records using 324 different models and kept 47 models for the next iteration, in which it processed 160 records and selected 7 models. Finally, it processed 240 records to select the single best model.

The error of the optimized model can be measured on the test set:

```
ŷtest  = predict(model,xtest)
ϵ      = relative_mean_error(ytest, ŷtest) # 0.07085
```

## 12.5.7 Model Interpretation and Feature Importance

With few exceptions, ML algorithms are often "black boxes" that may make good predictions, but whose predictions are very difficult to explain. Interpreting algorithms can relate to two different considerations. On the one hand, you may often be interested in understanding what is the general driver of the predictions—what are the most important ("significative") features. On the other hand, you may be interested in understanding what drives individual predictions—why the model made a particular prediction for a particular record.

BetaML provides a `FeatureRanker` model for the first type of analysis. `FeatureRanker` is a variable-ranking estimator that uses multiple metrics to rank variable importance. `FeatureRanker` helps you to determine the importance of features in the predictions of any black-box machine learning model (not necessarily from the BetaML suit), `FeatureRanker` internally uses cross-validation to assess the quality of the predictions, either by comparing the error made by the model with all features considered—with the error resulting from removing from the model one feature at a time (`metric="mda"`)—or by comparing the contribution of the variables to the variance of the predictions (`metric="sobol"`). Regardless of this setting, both measures are available by querying the model with `info(mod)`, this setting only determines which measure is used for the default ranking of the prediction output.

By default, `FeatureRanker` ranks variables (columns) in a single pass, without retraining on each one. However, it is possible to specify the model to use multiple passes (with the less important variables being removed on each pass). This helps to assess importance in the presence of highly correlated variables.

While the default strategy is to simply (temporarily) permute the "test" variable and predict the modified data set, it is possible to refit the model to be evaluated on each variable ("permute and relearn"), of course at a much higher computational cost. However, if the ML model to be evaluated supports ignoring variables during prediction (as BetaML tree models do), it is possible to specify the keyword argument for such an option in the prediction function of the target model and avoid refitting.

Take look again at the (x,y) data you used earlier:

```
x    = rand(1000,6)
y    = abs.([10*r[1]-r[2]+0.1*r[3]*r[1] + sqrt(r[6]*10) for r
in eachrow(x) ])
```

Now look at the importance of the features in predicting this dataset using a neural network model:

```
fr   = FeatureRanker(model=NeuralNetworkEstimator
(verbosity=NONE),nsplits=3,nrepeats=2,metric="mda")
rank = fit!(fr,x,y) # [4, 5, 3, 2, 6, 1]
```

You can obtain more information by querying the fitted model:

```
loss_by_col          = info(fr)["loss_by_col"]
sobol_by_col         = info(fr)["sobol_by_col"]
loss_by_col_sd       = info(fr)["loss_by_col_sd"]
sobol_by_col_sd      = info(fr)["sobol_by_col_sd"]
loss_fullmodel       = info(fr)["loss_all_cols"]
loss_fullmodel_sd    = info(fr)["loss_all_cols_sd"]
ntrials_per_metric   = info(fr)["ntrials_per_metric"]
```

The standard deviation and number of trials information allows you to plot the importance of the different variables with their confidence intervals, and to compare the error when all variables are considered (see Figure 12-10):

## CHAPTER 12  AI WITH JULIA

```
import Distributions: Normal
var_names=["x1","x2","x3","x4","x5","x6"]
bar(var_names[sortperm(loss_by_col)], loss_by_
col[sortperm(loss_by_col)],label="Loss by varᶜ",
permute=(:x,:y), yerror=quantile(Normal(0,1),0.975) .* (loss_
by_col_sd[sortperm(loss_by_col)]./sqrt(ntrials_per_metric)),
yrange=[0,0.2],legend=:bottomright, title="Feature importance")
vline!([loss_fullmodel], label="Loss with all
vars",linewidth=2)
vline!([loss_fullmodel-quantile(Normal(0,1),0.975) * loss_
fullmodel_sd/sqrt(ntrials_per_metric),
        loss_fullmodel+quantile(Normal(0,1),0.975) * loss_
        fullmodel_sd/sqrt(ntrials_per_metric),
], label=nothing,linecolor=:black,linestyle=:dot,linewidth=1)
```

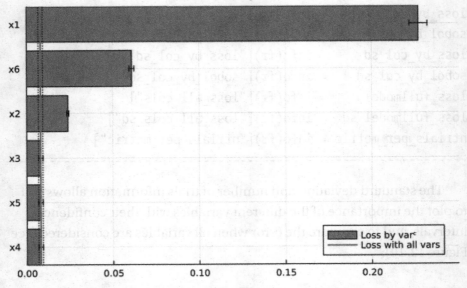

*Figure 12-10.* Loss increase when individual features are omitted

If you have paid attention to the additive function that has been used to generate y, you will notice that x4 and x5 do not contribute to the label at all, and are correctly reported as the "less important" variables. All x values are random numbers between 0 and 1, so if you cross-multiply them, you get very marginal contributions to y, and x3 follows in the rank order. x2 and x6 are next, and finally x1, which has both a component multiplied by 10 and another component crossed with x3, is considered the most important variable in predicting y.

## 12.6 Specialized AI Libraries in Julia

This section introduces some other ML libraries that are common in the Julia ecosystem or that I have used myself. The following list is a very partial set of the numerous packages that are available. Refer to the "AI" section of https://juliapackages.com/c/ai for a comprehensive list.

- **ML toolkits/pipelines**: MLJ.jl, ScikitLearn.jl
  MLJ is a particularly large project to enable different ML workflows. Like BetaML, it includes the algorithms as well as many "utility" and transformer models. In contrast to BetaML, each individual model is a wrapper around the model hosted in third-party packages (including BetaML). It uses the concepts of *machine*, where the model is coupled to the data, and *scientific types* of variables, which abstract from the computer implementation of particular data objects. Like MLJ, ScikitLearn is a wrapper for third-party models (many of which, as the name suggests, come from the Python scikit-learn library) and provides a syntax very similar to BetaML.

- **Neural networks**: `Flux.jl`
  Flux is the de facto standard for deep learning in Julia and includes high-level efficient implementations of many types of neural network architectures, including convolutional and recursive. It makes extensive use of automatic differentiation (AD), and any callable object can be a neural network layer.

- **Tree-based models**: `DecisionTrees.jl`
  Decision trees and random forests in DecisionTrees are very fast. However, they only work with ordered features and don't support missing data.

- **Clustering**: `Clustering.jl` and `GaussianMixtures.jl`
  Provide common clustering algorithms.

- **Missing imputation**: `Impute.jl`
  Provides only a basic set of imputation models, mainly along the interpolation method, whereas `Mice.jl` provides more methods, allows for multiple imputations, and can be combined with downstream statistical analysis to retain the uncertainty of the imputation in the overall analysis.

- **Variable meaning**: `ShapML.jl`
  ShapML provides sample-based Shapley scores related to the Sobol index used in `FeatureRanker`. This has the advantage that the index can be computed for individual observations.

# 12.7 EXERCISE 3: Predict the Values of Houses in Boston

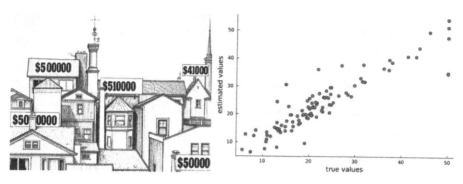

In this problem, you are given a dataset containing average house values in different Boston suburbs, together with the suburb characteristics (proportion of owner-occupied units built prior to 1940, index of accessibility to radial highways, etc.). Your task is to build a neural network model and train it to predict the average house value of each suburb.

The detailed attributes of the dataset are as follows:

1. CRIM: Per capita crime rate by town

2. ZN: Proportion of residential land zoned for lots over 25,000 sq. ft.

3. INDUS: Proportion of non-retail business acres per town

4. CHAS: Charles River dummy variable (= 1 if tract bounds river; 0 otherwise)

5. NOX: Nitric oxides concentration (parts per 10 million)

6. RM: Average number of rooms per dwelling
7. AGE: Proportion of owner-occupied units built prior to 1940
8. DIS: Weighted distances to five Boston employment centers
9. RAD: Index of accessibility to radial highways
10. TAX: Full-value property-tax rate per $10,000
11. PTRATIO: Pupil-teacher ratio by town
12. B: 1000(Bk - 0.63)^2, where Bk is the proportion of blacks by town
13. LSTAT: % lower status of the population
14. MEDV: Median value of owner-occupied homes in $1000's

You can find further information concerning this dataset file at `https://archive.ics.uci.edu/ml/machine-learning-databases/housing/housing.names`

Your predictions concern the median value (column 14 of the dataset).

**Skills used:** downloading and importing data from the Internet; designing and training a neural network for regression tasks using `BetaML`; using the additional `BetaML` functions `partition`, `scale`, and `meanRelError`; performing hyperparameters optimization and autotuning; interpreting your model, measuring the importance of the various features.

## 12.7.1 Instructions

The skeleton of the exercise presented next includes some code that is already set up, which you must complete by replacing [...] Write your code here with your own code. If you want to avoid typing, you can find this skeleton in the GitHub repository of the book (https://github.com/Apress/Julia-Quick-Syntax-Reference-2nd-ed), where you will also find its solution.

## 12.7.2 Skeleton

### 12.7.2.1 STEP 1: Set up the environment

Start by setting the working directory to the directory of this file and activating it. If you have the provided Manifest.toml file in the directory, just run Pkg.instantiate(); otherwise, manually add the packages Pipe.jl, HTTP.jl, CSV.jl, DataFrames.jl, Plots.jl, Distributions.jl, and BetaML.jl. Also, seed the random seed with the integer 123.

```
# [...] Write your code here
```

### 12.7.2.2 STEP 2: Load the packages

Load the packages Pipe.jl, HTTP.jl, CSV.jl, DataFrames.jl, Plots.jl, and BetaML.jl and import the functions quantile and the type Normal from the Distributions.jl package

```
# [...] Write your code here
```

## 12.7.2.3 STEP 3: Load the data

Load from the Internet or from a local file the input data into a data frame or matrix. You will need the CSV options header=false and ignorerepeated=true.

---
```
dataURL = "https://bit.ly/apress_julia_boston"
# [...] Write your code here
```
---

## 12.7.2.4 STEP 4: Create the feature matrix

Now create the X matrix of features (columns 1 to 13). Make sure that you have a 506×13 matrix (and not a DataFrame).

---
```
# [...] Write your code here
```
---

## 12.7.2.5 STEP 5: Build the label vector

Similarly, define Y as the 14th column of data.

---
```
# [...] Write your code here
```
---

## 12.7.2.6 STEP 6: Partition the data

Partition the (X,Y) data in (xtrain,xtest) and (ytrain,ytest), keeping 80% of the data for training and reserving 20% for testing. Keep the default option to shuffle the data, as the input data isn't shuffled.

```
# [...] Write your code here
```

### 12.7.2.7 STEP 7: Define the neural network architecture

Define a `NeuralNetworkEstimator` model with the following characteristics:

- Three dense layers with, respectively, 13, 20, and 1 nodes and activation function `relu`
- Cost function `squared_cost`
- Training options: 400 epochs and 6 records to be used on each batch

```
# [...] Write your code here
```

### 12.7.2.8 STEP 8: Train the model

Fit the model to the data using `ytrain` and a scaled version of `xtrain` (where all columns have zero mean and one standard deviation).

```
# [...] Write your code here
```

### 12.7.2.9 STEP 9: Predict the labels

Predict the training labels ŷtrain and the test labels ŷtest. Recall that you did the training on the scaled features!

```
# [...] Write your code here
```

### 12.7.2.10 STEP 10: Evaluate the model

Compute the train and test relative mean error using the function relative_mean_error.

```
# [...] Write your code here
```

### 12.7.2.11 STEP 11: Plot the errors and the estimated values vs. the true ones

Run the following commands to plot the average loss per epoch and the true vs. estimated test values:

```
plot(info(mynn)["loss_per_epoch"])
scatter(ytest, ŷtest, xlabel="true values", ylabel="estimated values", legend=nothing)
```

### 12.7.2.12 STEP 12: Hyperparameters tuning

Find the optimal model by testing the following hyperparameters: size of the middle layer, batch size, and number of epochs.

```
inner_layer_size_range = # [...] Write your code here
epoches_range          = # [...] Write your code here
bachsize_range         = # [...] Write your code here
```

## CHAPTER 12   AI WITH JULIA

If you are using the BetaML autotune mechanism, use the following line to build the range of the layers parameters to be used in the hpranges dictionary starting for a range defined in terms of the size of the inner layer:

```
layers_range = [[DenseLayer(13,i,f=relu),
DenseLayer(i,i,f=relu), DenseLayer(i,1,f=relu)] for i in inner_
layer_size_range]

# [...] Write your code here
```

### 12.7.2.13   STEP 13: Model interpretation

Determine the variable importance of the neural network model. The following vector provides the variable names of the features:

```
var_names = [
  "CRIM",    # per capita crime rate by town
  "ZN",      # proportion of residential land zoned for lots
             over 25,000 sq.ft.
  "INDUS",   # proportion of non-retail business acres per town
  "CHAS",    # Charles River dummy variable  (= 1 if tract
             bounds river; 0 otherwise)
  "NOX",     # nitric oxides concentration (parts per 10
             million)
  "RM",      # average number of rooms per dwelling
  "AGE",     # proportion of owner-occupied units built
             prior to 1940
  "DIS",     # weighted distances to five Boston
             employment centers
```

```
    "RAD",     # index of accessibility to radial highways
    "TAX",     # full-value property-tax rate per $10,000
    "PTRATIO", # pupil-teacher ratio by town
    "B",       # 1000(Bk - 0.63)^2 where Bk is the proportion of
               blacks by town
    "LSTAT",   # % lower status of the population
]
```

Which are the most important variables to correctly predict the average house value ?

```
# [...] Write your code here
```

### 12.7.3 Results

You should arrive to a model that is able to predict the average house value, on unseen data, with an error around 12% to 13%. The feature importance analysis should highlight the influence of the characteristics of the local population (richness and black status) on the house value.

### 12.7.4 Possible Variations

You may want to consider other NN architectures or try other ML models, such as RandomForests or those based on Perceptron. Which are better for this particular task? You may want to try other options of the FeatureRanker model, such as calculating the rank using the Sobol metric, re-estimating the model at each column, or calculating the rank iteratively, dropping the less important feature at each iteration.

# CHAPTER 13

# Utilities

The following third-party packages are covered in this chapter:

| Weave.jl | https://github.com/mpastell/weave.jl | v0.10.12 |
|---|---|---|
| ZipFile.jl | https://github.com/fhs/ZipFile.jl | v0.10.1 |

This chapter also mentions a few other Julia packages that are of more general use.

## 13.1 Weave.jl for Dynamic Documents

Weave.jl (https://github.com/mpastell/weave.jl) allows developers to create "dynamic documents" in which the code that produces the "results" is embedded directly in the document. The overall structure of the document is a *Markdown* document, while the Julia code is inserted as *code cells* within the main document.

In many situations, this is very useful, as it allows the logic from assumptions to results, and the presentation and discussion of those results, to be presented in the same document.

A Weave document is similar to a Jupyter Notebook document in that they both can include code and Markdown-based documentation. The main advantage of Weave is that the blocks of code tu run (called _cells_ in Jupyter) are not defined _a priori_, when you write the code. Instead you can still run the part of the code you need, at least when you use Weave when developing in VS Code. This regardless of how you have organized the cells in the document.

The following is an example of a Weave *Julia Markdown* document, which you write in a file with the .jmd extension (e.g., test_weave.jmd):

---

```
write("test_weave.jmd","""
    ---
    title :          Test of a document with embedded Julia code
                     and citations
    date :           18 August 2024
    bibliography: biblio.bib
    ---


    ```{julia;eval=true,echo=false,results="hidden"}
    # (leave two rows from the document headers above)
    # This code is hidden in the output - both the code and
      its output.
    # You can use to initialize the script, here for example to
      install the packages
    # that the script requires
    using Pkg
    Pkg.add(["Plots","DataFrames"])
    ```

    # Section 1
    Weave.jl, announced in @Pastell:2017, is a scientific
    report generator/literate programming tool for Julia
    developed by Matti Pastell, resembling  Knitr for R [see
    @Xie:2015].
```

## Subsection 1.1

This should print a plot. Note that, with `echo=false`, you are not rendering the source code in the final PDF:

```
{julia;echo=false}
using Plots
plot(sin, -2pi, pi, label="sine function")
```

Here instead you will render in the PDF both the script source code and its output:

```
{julia;}
using DataFrames
df = DataFrame(
        colour = ["green","blue","white","green","green"],
        shape  = ["circle", "triangle", "square",
                  "square","circle"],
        border = ["dotted", "line", "line", "line",
                  "dotted"],
        area   = [1.1, 2.3, 3.1, missing, 5.2]
    )
df
```

Note also that you can refer to variables defined in previous chunks (or "cells", following Jupyter terminology):

```
{julia;}
df.colour
```

CHAPTER 13   UTILITIES

```
    ### Subsubsection

    For a much more complete example see the [Weave
    documentation](http://weavejl.mpastell.com/stable/).

    # References
    """)
```

---

The document above is a Weave _Julia Markdown_ (.jmd) document. The `biblio.bib` referenced in the Wave file is a standard BibTeX file containing the necessary citations:

---

```
write("biblio.bib","""
@article{  Pastell:2017,
  author   = {Pastell, Matti},
  title    = {Weave.jl: Scientific Reports Using Julia},
  journal  = {Journal of Open Source Software},
  vol      = {2},
  issue    = {11},
  year     = {2017},
  doi      = {10.21105/joss.00204}
}
@Book{     Xie:2015,
  title      = {Dynamic Documents with R and Knitr.},
  publisher  = {Chapman and Hall/CRC},
  year       = {2015},
  author     = {Yihui Xie},
  edition    = {2nd ed},
  url        = {http://yihui.name/knitr/},
}
""")
```

---

If you are in VS Code, keep test_weave.jmd open in the main panel and "play" with the Julia code. When you are satisfied, you can compile the document by running the following commands from the Julia console in VS Code:

```
julia> using Weave;
julia> weave("test_weave.jmd", out_path = :pwd)
julia> weave("test_weave.jmd", out_path = :pwd, doctype = "pandoc2pdf")
```

Assuming that the necessary tools are present in the system[1], the first command would produce an HTML document, while the second would produce the PDF shown in Figure 13-1.

---

[1] On Ubuntu Linux (but probably on other systems too), Weave requires Pandora >= 1.20 and LaTeX (texlive-xetex) to be installed on the system. If you are using Ubuntu, the version of Pandora in the official repositories may be too old. Use the deb from https://github.com/jgm/pandoc/releases/latest instead.

# CHAPTER 13  UTILITIES

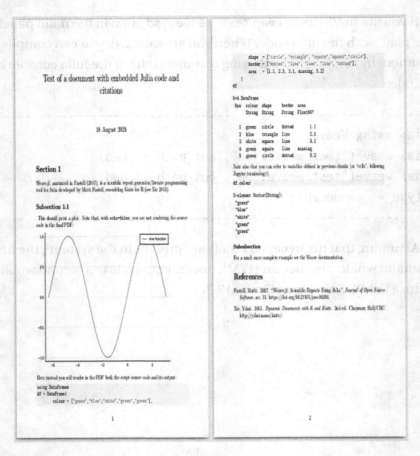

***Figure 13-1.*** *Weave compiled PDF document*

As an alternative to Weave.jl, you can use Literate.jl (https://github.com/fredrikekre/Literate.jl) to write the original document as Julia script, with the Markdown chunks embedded using a special comment syntax. Literate.jl transforms the Julia code with embedded Markdown into Markdown files with embedded Julia code chunks, using similar options that Weave uses to control the parsing and display of the code and output. Another option is to use Quarto, which now supports Julia. As a rule of thumb, I prefer to write my original documents in Julia with embedded Markdown when the main interest is in the code itself and the Markdown is used just for comments, documentation, or a way of providing structure to the document; I prefer to write in Markdown with embedded code when the main interest is to communicate something I am using the code to accomplish.

## 13.2 ZipFile

If you have worked with compressed files, you likely have dealt with ZIP archives. ZipFile.jl (https://github.com/fhs/ZipFile.jl) provides a library that enables you to manage ZIP archives easily.

### 13.2.1 Writing a Zip Archive

To write a ZIP archive, you first open the ZIP file that serves as the container. Then you add the files you need to compress and write them into the ZIP file. As the final step, close the ZIP archive, which is important because at this point the data is flushed to disk (at least for small files) and the ZIP file is finalized.

```
zf = ZipFile.Writer("example.zip")
f1 = ZipFile.addfile(zf, "file1.txt", method=ZipFile.Deflate)
write(f1, "Hello world1!\n")
write(f1, "Hello world1 again!\n")
f2 = ZipFile.addfile(zf, "dir1/file2.txt", method=ZipFile.Deflate)
write(f2, "Hello world2!\n")
close(zf) # Important!
```

The package does not export `Writer()` or `addfile()`, so you need to prepend them to the package name as shown in the example. Also, when adding a file, you must specify `method=ZipFile.Deflate`. Otherwise, the file will be stored uncompressed in the ZIP archive, and that is not what you want in most situations.

In the previous example, you used two different objects for the two files. Note that a bug in the library prevents you from adding all the files and then writing to them in a second step. For instance, this would not work:

```
zf = ZipFile.Writer("example.zip")
f1 = ZipFile.addfile(zf, "file1.txt", method=ZipFile.Deflate)
f2 = ZipFile.addfile(zf, "dir1/file2.txt", method=ZipFile.Deflate)
write(f1, "Hello world1!\n") # Error !
write(f1, "Hello world1 again!\n")
write(f2, "Hello world2!\n")
close(zf) # Important!
```

## 13.2.2 Reading from a Zip Archive

The process for reading from a ZIP archive is similar. First, you open the archive in read mode with `ZipFile.Reader` (again, you need to use the function name with the package name in front), and then you can loop over its `files` property and read from it, using either `eachline` or `read`.

You don't need to close each file, but you do need to close the ZIP archive:

```
zf = ZipFile.Reader("example.zip");
for f in zf.files
    println("*** $(f.name) ***")
    for ln in eachline(f) # Alternative: read(f,String) to read
                            the whole file
```

```
        println(ln)
    end
end
close(zf) # Important!
```
---

You can also get other information about the file by looking at its name, method (Store or Deflate), uncompressedsize, and compressedsize properties (e.g., f.name).

To get information about the whole ZIP archive, use show(zf) instead:

---
```
julia> zf = ZipFile.Reader("example.zip");
julia> show(zf)
ZipFile.Reader for IOStream(<file 11_example.zip>) containing 2 files:
uncompressedsize method   mtime                name
             34 Deflate 2019-07-18 12-06 file1.txt
             14 Deflate 2019-07-18 12-06 dir1/file2.txt
julia> close(zf)
```
---

# Index

## A

Abstract syntax tree (AST), 116, 117
Abstract types, 86
    keyword abstract type, 79
    multiple-dispatch mechanism, 81
    multiple inheritance, 79
    MyOwnAbstractType, 81
    MyOwnGeneric AbstractType, 80
Activation function, 173, 298–300, 302
AD, see Automatic differentiation (AD)
add PackageName, 18
AIC, see Akaike Information Criterion (AIC)
Akaike Information Criterion (AIC), 281
Algebraic Modeling Language (AML), 230, 231
allowmissing(array), 53, 199
an_exported_var, 15
AML, see Algebraic Modeling Language (AML)
Anonymous functions, 31, 62, 69, 71
API, see Application programming interface (API)
Application programming interface (API), 11, 21, 66, 187, 190, 201, 241, 252, 255, 268, 271, 283
Arguments function, 63–65
Array of arrays, 32, 289
Arrays (lists)
    bounding limits, 29
    column vector (one-dimensional array), 28
    empty (zero-element) arrays, 28
    functions, 30–32
    n-elements array of identical j elements, 28
    matrices, 27
    multidimensional arrays, 32–37
    n-elements array of random numbers, 28
    n-elements array whose content is garbage, 28
    n-elements ones array, 28
    n-elements zeros array, 28
    N-dimensional mutable containers, 27
    nested arrays, 32–37

# INDEX

Arrays (lists) (*cont.*)
    row vector, 28
    square brackets, 29
    store heterogeneous types, 28
    Union keyword, 28
    vectors, 27
Array{T,n}, 186
AST, *see* Abstract syntax tree (AST)
atexit(f) function, 72
AutoEncoder, 285–288, 295
Automatic differentiation (AD), 302
Autonomous robots, 266, 267
Autotune, 293, 321–323, 325

## B

Base.show function, 109
Bayesian Information Criterion (BIC), 281, 315, 316
Benchmarking, 152–155
BenchmarkTools.jl package, 154
BetaML functions, 272
BetaML.jl package, 200, 201, 265, 268, 331
BetaML models, 273, 321
BetaML NeuralNetwork Estimator, 284
BetaML Toolkit, 268
    data and workflow processing, 270
    fit! function, 272
    hyperparameters, parameters and options, 274
    inverse_predict, 274
    manage stochasticity, 273–274
    reset!(mod), 274
    supervised models, 268
    unsupervised models, 269
    verbosity, 271
BIC, *see* Bayesian Information Criterion (BIC)
Bits type, 87
Black boxes, 323, 324
Boolean array, 35, 199
Boolean selection, 35, 191, 192, 194, 261
Boolean values, 24, 31, 44, 192, 195
broadcast() function, 70
Broadcast function, 64, 70

## C

Call by reference/call by value, 68, 69
Call-by-sharing, 68
CategoricalArrays.jl package, 185, 195, 197
C compiler, 127
C++ code, 131, 139
C++ function, 130
Char type, 24
for c in eachcol(df) function, 190
Classification models, 298, 310–314
Clustering, 267, 305–308
Clustering.jl, 328
Clustering models, 314–316
combine function, 206, 207

# INDEX

Comma-separated value (CSV) files, 102–103
Composite type, 73, 82, 86, 87, 161
Composition, 80, 82, 83, 85, 173
Concatenate strings, 26
Concrete type, 79, 86, 160
CondaPkg.toml file, 138
Conda Python environment, 141
Conda.update command, 146
Conditional breakpoint, 167
Conditional statements
    if blocks, 60, 61
    ternary operator, 60, 61
ConfusionMatrix model, 312
Constructors, 76–78, 84
Container/collection, 87
Continuous distributions, 254
copy() or deepcopy(), 47
C programs, 152
cross_validation, 321
cross_validation(f,data, sampler), 320
CSV, *see* Comma-separated value (CSV) files
CSV.jl package, 18, 103, 108, 257, 258
CSV package, 233
CSV.read, 103
CSV.Tables.matrix, 103
CSV.Tables.table(my_matrix) function, 111
Current directory, 20
CxxWrap, 128–131, 133
Cython, 4

## D

DataFrames, 37, 103, 104, 111, 113, 145, 187, 189, 194, 195, 205, 207
DataFrame export
    CSV, Excel or ODS format, 209
    Dict, 210, 211
    HDF5.jl package, 211, 212
    Matrix constructor, 210
DataFrames.jl package, 186
    categorical data, 197, 198
    creating dataframe/loading data, 187, 188
    data filtering, 190–193
    editing data, 193, 194
    editing the structure, 195–197
    export (*see* Dataframe export)
    installing and importing library, 187
    managing missing values, 199–201
    pivoting data, 202–204
    split-apply-combine strategy, 206–209
    structure analysis using functions, 189, 190
    two-dimensional (tabular) data, 187
DataFrameMeta.jl package, 190, 192
DataFrames package, 150, 199, 204, 206
Data partitioning, 289

# INDEX

Data preprocessing
    data partitioning, 289
    dimensionality
        reduction, 285–288
    encoding categorical
        data, 275–277
    GaussianMixtureImputer,
        280, 281
    GeneralImputer, 283, 284
    RandomForestImputer, 282
    scaling, 277–279
    SimpleImputer, 280
    "tidy", 275
Data-storage formats, 114
Data types and structures
    arrays (*see* Arrays (lists))
    dates and times (*see* Dates
        and times)
    dictionaries, 39–41
    memory and copy issues, 47–50
    named tuples, 39
    random numbers, 50–52
    sets, 41
    simple types (non-containers)
        basic mathematical
            operations, 25
        Boolean values, 24
        default floating-point
            type, 25
        default integer type, 25
    strings
        concatenation, 26
        integers/floats, 26
        Julia, 25
        operations, 26
        single row, 26
    tuples, 38, 39
    various notes, data types
        const keyword, 53
        referenced object, 53
        variable references, 54
Dates, 47
Dates and times
    DateTime, 42
    date-time arithmetic, 45–47
    date/time object ("Input"),
        creation, 42–44
    extract information, date/time
        object ("output"), 44, 45
    formatters, 43
    periods, 45–47
DateTime, 42
Debug Console, 168
Debugging, 162, 165–168
"Deep" copy, 47
Default positional constructor, 78
DelimitedFiles.jl, 102
DelimitedFiles.writedlm
    function, 111
describe(df) function, 189
destination_sink, 103
@df macro, 224
Dictionaries, 39–41
disallowmissing(array), 53, 199
Discrete distributions, 253, 255, 256
Distance-based clusterers, 305
Distributions.jl package, 51,
    226, 331

API, 255
continuous distributions, 254
discrete distributions, 253
object definition, 252
supported distributions, 253
do block, 71, 112
Domain-specific languages (DSLs), 116
dot notation, 39, 64, 70, 106, 140
dot (.) operator, 53, 84
DSLs, *see* Domain-specific languages (DSLs)

# E

Efficient Code
    VS Code, 157
[eltype(c) for c in eachcol(df)] function, 190
Environment directory, 20
Environments, 17–20
ExcelFiles, 104
Expressions
    AST, 117
    colon prefix operator, 119
    create by parsing the computer code, 117, 118
    equal sign, 118
    evaluate symbols, 119–121
    Exp constructor, tree, 119
    Julia, 117
    objects, 117
    parse a string, 118
    quote block, 119

# F

FeatureRanker model, 324, 336
FeatureRanker ranks variables, 324
Fibonacci number, 153
Fibonacci function, 153
File system functions, 98, 99
first(df,n) function, 189
Float64, 25, 27, 53, 113, 128, 175
5-fold cross-validation, 319
Foo module, 14, 15
Foo.jl, 15
Forest growth model
    computing age class, 261
    computing standard error, 263
    computing volumes, 259, 260
    data filtering, 261
    defining fit model, 262
    downloading and importing data, 257
    environment setup, 257
    filtering unused information, 259
    initialParameters, 262
    instructions, 257
    join datasets, 261
    loading data, 258
    loading packages, 258
    logisticModelVec, 263
    model differentiation, 264
    plot fitted model, 263
    scatter chart, 264
    trees dataframe aggregation, 260

# INDEX

Functions
    anonymous, 69
    arguments, 63–65
    broadcast, 70
    call by reference/call by value, 68, 69
    convention, 68
    definition, 62
    multiple-dispatch (a.k.a. polymorphism), 66, 67
    nested, 62
    as objects, 67
    return value, 66
    rules, 63
    templates, 67

## G

GaussianMixtureClusterer, 305, 306, 315
GaussianMixtureClustering, 315
GaussianMixtureImputer model, 200, 280, 281, 283
GaussianMixtures.jl, 328
GeneralImputer, 201, 283–284
Generative (Gaussian) Mixture Model (GMM), 280
Generic loop, 122
getUserInput function, 101
Global Julia startup file, 9
Global variables, 53, 56, 57, 160, 168, 180
GMM, *see* Generative (Gaussian) Mixture Model (GMM)

GPU, *see* Graphical processing unit (GPU)
Graphical processing unit (GPU), 171
    array, 174
    computation, 174
    computations, 176
    CPU computing, 176
    kernels, 173, 174
    packages, 176
    programming, 172, 177
    vendors, 175
groupby function, 209

## H

HDF5.jl package, 211, 212
Help system, 21
High-level languages, 48, 62, 76
HiGHS.jl, 233
Hyperparameters, 271, 274
Hyperparameters evaluation, 317–318

## I

IDEs, *see* Integrated development environments (IDEs)
IJulia, 7
IJulia kernel, 7
IJulia; notebook(), 7
IJulia notebooks, 21
Image recognition, 266
Immutable, tuples, 37, 38

Immutable type, 86
Imputation module, 200
Impute.jl, 328
include("myScript.jl"), 100
IndexedTables
    creating (NDSParse), 213, 214
    editing/adding values, 215
    flavors, 212
    JuliaDB ecosystem, 212
    row filtering, 215
    table-like data structures, 212
IndexedTables.jl package, 186
InexactError, 34, 46, 79
Inheritance, 79, 80, 82, 83, 85
Input/output (I/O), Julia
    file system functions, 98, 99
    IOStream object, 97
    other specialized IO, 114
    reading (input) (*see* Reading (input))
    user's terminal, 97
    web resources, 97
    writing (output) (*see* Writing (output))
Int64, 25
Integer divisions, 25
Integrated development environments (IDEs), 5, 6, 8
Interpolation, 14, 27, 28
Introspection, 5, 163–165
IOStream object, 97, 110
Ipopt.jl, 233

## J

JavaScript Object Notation (JSON), 104–106, 114
JSON, *see* JavaScript Object Notation (JSON)
JSON3.jl package, 105, 114
JSON3.Object, 106
JSON.jl, 105
Julia, 125, 152, 155, 160, 164, 169
    ccall function, 127
    C code, 126
    C function, 128
    C library, 126
    composition, 82, 85
    computational advantages, 5
    conditional statements
        if blocks, 60, 61
        ternary operator, 60, 61
    cpp_hello function, 129
    C++ source, 129
    current directory *vs.* environment, 20
    C++ workflow, 128
    CxxWrap, 130, 133
    definitions, common Julia terms, 86
    developer, 3
    development, 125
    do blocks, 71
    environments, 17–20
    exit, 72
    exit(exit_code), 72
    features, 5

INDEX

Julia (*cont.*)
   functionality, 126
   functions (*see* Functions)
   general-purpose programming language, 229
   help system, 21
   improve runtime execution, dynamic languages, 4
   improve trade-off, 3
   inheritance, 82, 83, 85
   installation, 6–8
   JIT compilers, 4
   language features, 4
   languages, 125, 126
   miscellaneous syntax elements, 10–11
   missingness concepts, 53
      missing, 52
      NaN, 53
      Nothing, 52
   modules, 14–15
   multiple dispatch, 84
   package manager, 12–13
   packages, 16–17
   parametric types, 24
   R code, 143
   registered packages, 12
   repeated iteration
      list comprehension, 57–60
      maps, 57–60
      for and while loops, 57–60
   run, 8–9
   runtime performances, traditional high-level dynamic languages, 5
   shadow costs, new language, 4, 5
   soft scope, 56
   specialization, 85
   specialized type, 84
   typical flow-control constructs, 55
   unregistered packages, 12
   Variable Scope, 56
   weak relation, 85
Julia Base, 53
JuliaCall package, 139, 147, 148
Julia DataFrame object, 150
Julia development environments, 11
Julia ecosystem, 327
julia_eval() function, 149
Julia functions, 68, 126, 140, 149
JuliaInterpreter.jl package, 151, 165
julia myScript.jl, 8, 100
Julia packages, 114, 134, 141, 142, 150, 158, 231, 233, 268, 337
JuliaPackages.com, 327
Julia program, 135, 152
Julia programming language, 268
Julia REPL, 144, 159
Julia RNG, 51
Juliaup, 6, 7
Julia VS Code extension, 165

354

JuMP
  AML, 230
  general settings, 231
  linear problem (*see* Transport problem)
  nonlinear problems (*see* Nonlinear problems)
  third generation, 231

# K

Keyword arguments, 63, 222, 223, 227
K-fold cross-validation, 319–321
KMeansClusterer, 269, 305, 306
KMedoidsClusterer, 269, 305, 306
kwarg, 77
@kwdef keyword, 76

# L

Lagrangian, 246
Lagrangian multiplier, 246
Language Integrated Query (LINQ), 209
last(df,n) function, 189
LaTeX-like syntax, 11
Learning rate, 301
Library package, 179
LINQ, *see* Language Integrated Query (LINQ)
List comprehension, 60, 61
Literate.jl, 342
Local Julia startup file, 9

Logical operators, 61
Long format, 202
Loop matrices, 161
LsqFit.jl package
  fitting model, 250
  loading libraries and defining model, 249
  parameters, 250
  retrieving parameters, 251

# M

Macros, 116
  definition, 122
  invocation, 123
  Julia, 121
  string, 124
Main.Foo, 15
Manifest.toml, 17, 19
map, 60
Markdown document, 337
Matrix constructor, 210
Matrix data, 161
MatrixTable, 111
Merging/joining/copying datasets, 196, 197
Metaprogramming
  AST, 116
  C++ macro, 115
  code substitution, 116
  DSLs, 116
  for loops, 115
  programmers, 115
  programming languages, 115

# INDEX

method=ZipFile.Deflate, 343
MINLP, *see* Mixed-Integer
    Nonlinear
    Programming (MINLP)
MinMaxScaler, 277
Miscellaneous Syntax
    Elements, 10–11
Missing imputation, 328
Missings.jl package, 53
Missing values
    management, 199–201
Mixed-Integer Nonlinear
    Programming (MINLP), 242
ML toolkits/pipelines, 327
Model constructor, 236
Modules, 14–15
Multidimensional arrays
    Boolean selection, 35
    list comprehension, 33
    one-dimensional arrays, 32, 33
Multiline comments, 10
Multiple conditions, 61
Multiple-dispatch (a.k.a.
    polymorphism), 66, 67
Mutable types, 86
MyOwnGenericAbstractType, 80
MyOwnType, 77

# N

Name binding, 47
Named tuples, 39, 41
names(df) function, 189
Name tuples, 41

Natural exponential expressions, 25
Natural language processing, 266
n-dimensional arrays
    elements, 34
    functions, 35–37
ndsparse function, 213, 214
Nested arrays
    double square brackets, 34
Neural network model
    house values in different Boston
        suburbs, 329–336
Neural networks, 328
    deep neural network, 296
    in BetaML, 298
        ConvLayer, 300
        DenseLayer, 299
        DenseNoBiasLayer, 299
        GroupedLayer, 299
        one-dimensional regression
            models, 298
        PoolingLayer, 300
        ReplicatorLayer, 299
        ReshaperLayer, 299
        ScalarFunctionLayer, 299
        schema, 297
        single-neuron schema, 297
        VectorFunctionLayer, 299
    example, 303–304
    neural network
        training, 300–303
    specialized hardware, 296
    supervised models, 295
Nonlinear problems
    declaring model variables, 242

# INDEX

importing libraries, 242
problem definition, 241
resolving model and results, 243, 244
nonmissingtype(Union {T,Missing}), 53
Nottingham Forest, 105
Numba, 4
Numpy array, 140

## O

Object initialization, 77
OdsIO.jl package, 104
ods_write(filename,data), 113
One-hot encoding, 275
OOPMacro.jl package, 85
open() function, 110
Optimization problems
    definition, 230
    software tools, 230
    solver-based approaches, 230

## P

Package commands
    add git@github.com:userName/ pkgName.jl.git, 13
    add pkgName, 13
    add pkgName#branchName, 13
    add pkgName#main, 13
    add pkgName#vX.Y.Z, 13
    free pkgName, 13
    rm pkgName, 13
    status, 13
    update, 13
Package Manager, 12–13
PackageName, 16, 17
PackageName:package_function, 17
Packages, 17
Parallelization, 171, 181, 182
Parametric type, 86
PCAEncoder, 285
PegasosClassifier, 290
Perceptron-like classifiers, 290–293
Periods and date-time arithmetic, 45–47
@pipe macro, 258
pipe operator, 150, 216, 217
Pipe.jl package, 108, 217
Pipe operator, 209
Pivoting
    definition, 203
    stacking columns, 204
    unstacking, 204
Pkg module, 12
plot() function, 16, 220–224
PlotlyJS.js, 219
Plots.jl, 16
    backends, 218–220
    combining multiple plots, 227, 228
    plot function, 220–223
    plotting densities and distributions, 226
    plotting from Dataframes, 224–226
    saving plot, 228

Plotting, 218
pmap function, 181
Polymorphism, 66
Positional arguments, 63
Position-based object instantiation, 77
Predefined type, 87
Primitive type, 86
 current limitation, 74
 definition, 74
Primitive types, 73
print() outputs, 109
println, 14
Profile data, 155
Project.toml, 17, 19
Pseudo-random numbers, 50
PyJuliaPkg, 142
PyPy, 3
Python-based indexes, 138
Python API, 138
PythonCall command, 134, 135, 137, 145
Python code, 134
 dot notation, 140
 ezodf module, 138
 features, 134
 functions, 136
 Julia, 136
 Julia code, 134
 Julia functions, 140
Python ezodf module, 137
Python programmers, 139
Python programming language, 4
Python script, 139

## Q

Quarto, 342
Query.jl package, 190, 193
Quote Block, 119

## R

RandomForestImputer, 200, 282–283
uses Random.GLOBAL_RNG, 51
Random number generator (RNG), 51, 273
Random numbers, 50–52
R code
 RCall package, 143
 RCall.jl, 143
 R_HOME, 144
readdlm output, 103
Reading (input)
 access web resources, 106–108
 file, 101
 import data from Excel, 104
 import data from JSON, 104–106
 read the whole file, single operation
  CSV files, 102, 103
  file line by line, 102
  terminal, 100, 101
Reduction, 208
Ref() function, 70
Registered packages, 12
Regression, 266
Regression models, 308–310
Regularization techniques, 317

Reinforcement learning, 267
Return value, 66
R functions, 145
for r in eachrow(df)
    function, 190
Row filtering, 215
RNG, *see* Random number
    generator (RNG)
R package, 146

# S

Split-apply-combine
    strategy, 206–209
savefig(plot_object,
    filename), 228
Scalar, 23
Schelling segregation model
  Agent class, 88
  generality *vs.* specificity, 88
  gridded space, 88
  possible variations, 95
  results, 95
  simulation algorithm, 88
  skeleton
    define Agent and Env
        classes, 89
    initialize the simulation with
        given parameters, 94
    main functions,
        algorithm, 91
    model run, 95
    set the parameters, specific
        simulation to run, 93

  set up the environment, 89
  utility functions, 90
  skills uses, 88
  social sciences, 87
Scope, 56
Serde.jl, 105
set_optimizer(model,
    optimizer), 236
Sets, 41, 233
seval() function, 139
SGD, *see* Stochastic Gradient
    Descent (SGD)
Stochastic Gradient Descent
    (SGD), 301
show(df) function, 189
SimpleImputer, 200, 280
SimpleTraits.jl package, 85
Simulation algorithm, 88
Singleton, 86
size(df) function, 190
skipmissing(array), 53
Sorting, 195
Specialized AI libraries, Julia
    ecosystem, 327–328
Splat operator, 65
StableRNGs.jl, 52
stack(df,[cols]) function, 204
Stacking columns, 204
StandardScaler, 277
startup.jl file, 9
StatsPlots.jl package, 224
STD objects, 131
String macros, 124
string(mysymbol), 117

# INDEX

Structures
    composite types, 73
    definition, 75, 76
SuccessiveHalvingSearch, 322
Supervised learning, 266
Symbol() function, 116
Symbols, 116, 117
SymPy
    computer algebra system, 244
    creating and manipulating expressions, 246, 247
    documentation, 244
    loading libraries, 245
    retrieving numerical values, 248
    system of equation, 247
SymPyPythonCall.jl, 244
Syntax, 41

## T

Templates, 76
Templates (type parameterization), 67
Ternary operator, 61
test_weave.jmd, 341
The :: operator, 74
Third-party packages, 17, 21, 23, 97, 125, 151, 171, 185
Threads standard library, 176
"Tidy" format, 275
to_dict, 211
Traditional object-oriented languages, 66
Trait, 87

Transport problem
    importing libraries, 233
    model constraints declaration, 237
    model declaration, 236
    model, human-readable visualization, 238
    model objectives declaration, 237
    model resolution, 238
    model variables declaration, 236
    parameter definition, 234
    problem definition, 232
    results visualization, 239, 240
    sets definition, 233
Tree-based models, 293–295, 328
Tuples, 38, 39
TypeError, 103
Type-related functions, 85
Type stability, 159
    parameters, 158
Type-unstable functions, 158
Type *vs.* structure, 73

## U

Unicode symbols, 11, 25
unique(df.fieldName) function, 190
Unregistered packages, 12
Unstacking, 204–205
Unsupervised learning, 267, 268
User-defined primitive type, 74
Utility function, 90, 241, 248

# V

@variable macro, 236
Variable meaning, 328
Variable references, 54
Variables
    global variable a, 56
    block or function, 56
    global, 56
Variable Scope, 56
Vector{T}, 27
Verbosity, 271
versioninfo(), 9
VS Code, 6, 15, 21, 165
VS Code documentation, 169
VS Code Julia extension, 6
VS Code profiler, 157

# W, X

Weave, 337, 340
Weave Julia Markdown document, 338
Weave Julia Markdown (.jmd) document, 340
Weave compiled PDF document, 342
Weave.jl, 337, 342

for and while Loop, 57, 59
Wide format, 202
workspace() function, 72
write() outputs, 109
Writer(), 343
Writing (output)
    export to CSV, 111
    export to Excel and ODS files, 112, 113
    export to JSON, 114
    file, 110, 111
    terminal, 108, 109

# Y

YAML.jl package, 105
Yype-unstable function, 158

# Z

Zero-indexing standard, 11
ZipFile.jl
    managing zip archives, 343
    reading from zip archives, 344, 345
    writing zip archives, 343, 344
ZipFile.Reader, 344

Printed in the United States
by Baker & Taylor Publisher Services